Atlases of Clinical Nuclear Medicine

Series Editor: Douglas Van Nostrand

Atlases of Clinical Nuclear Medicine

Series Editor: Douglas Van Nostrand

Selected Atlases of Gastrointestinal Scintigraphy
Edited by
Harvey A. Ziessman and Douglas Van Nostrand

Selected Atlases of Bone Scintigraphy
Edited by
Sue H. Abreu, Harvey A. Ziessman, and Douglas Van Nostrand

Selected Atlases of Cardiovascular Nuclear Medicine
Edited by
Sue H. Abreu and Douglas Van Nostrand

Selected Atlases of Renal Scintigraphy
George N. Sfakianakis

Harvey A. Ziessman
Douglas Van Nostrand
Editors

Selected Atlases of Gastrointestinal Scintigraphy

With 132 Figures in 278 Parts

Springer-Verlag
New York Berlin Heidelberg London Paris
Tokyo Hong Kong Barcelona Budapest

Harvey A. Ziessman, MD
Associate Professor of Radiology, Division of Nuclear Medicine,
Georgetown University Hospital, Washington, D.C. 20007, USA

Douglas Van Nostrand, MD
Director, Nuclear Medicine Department,
Good Samaritan Hospital,
Baltimore, MD 21239, USA
and
Clinical Professor of Radiology and Nuclear Medicine, Uniformed
Services University of Health Sciences, Bethesda, MD 20814, USA

Library of Congress Cataloging-in-Publication Data
Selected atlases of gastrointestinal scintigraphy / Harvey A.
  Ziessman, Douglas Van Nostrand, editors.
        p.   cm.
    Includes bibliographical references and index.
    ISBN-13: 978-1-4612-7673-9   e-ISBN-13: 978-1-4612-2794-6
    DOI: 10.1007/978-1-4612-2794-6
      1. Gastrointestinal system — Radionuclide imaging — Atlases.
  I. Ziessman, Harvey A.   II. Van Nostrand, Douglas.
      [DNLM: 1. Gastrointestinal Diseases — radionuclide imaging-
  -atlases.   2. Gastrointestinal System — radionuclide imaging-
  -atlases.   WI 17 S464]
  RC804.R27S45   1991
  616.3'07575 — dc20
  DNLM/DLC
  for Library of Congress   91-4853

Printed on acid-free paper.
© 1992 Springer-Verlag New York, Inc.
Softcover reprint of the hardcover 1st edition 1992

Typeset by Bytheway Typesetting Services, Norwich, NY.

9 8 7 6 5 4 3 2 1

Series Preface

Atlases of Clinical Nuclear Medicine will be a sequence of approximately three to five moderately sized and priced books to be published periodically every one to two years. The series will cover a wide range of subjects, and in each volume typically three to five extensive atlases of different imaging procedures or specific aspects of an imaging procedure will be presented. In some volumes, all chapters will cover a specific organ system, such as gastrointestinal scintigraphy or cardiac nuclear medicine; and some volumes will have chapters from several organ systems. The topics of the specific chapters in the atlases will usually include several chapters of current interest and one or two chapters of less frequently performed procedures. However, all of the chapters will be typically directed toward the clinical practice of nuclear medicine.

The purpose of this series is to bring to the reader selected atlases of nuclear medicine, which (1) have never been published before, (2) are more extensive that those previously published, or (3) are more current than those previously published.

The series will be of value to the practicing physician and radiologist as well as the resident learning clinical nuclear medicine. The nuclear medicine physician or radiologist will find these atlases a source of practical information for procedures that he or she already performs as well as for specific aspects of a procedure that he or she is only occasionally called upon to perform and interpret. For the physician learning nuclear medicine, these atlases will be an excellent training tool and source of information. Teaching points are emphasized. In addition, other physicians from associated specialties such as gastroenterology, orthopedic surgery, and cardiology will find individual volumes valuable.

The typical atlas will feature an introductory text followed by a gallery of images. In the introductory text, such items as technique (imaging procedure, computer acquisition analysis), physiologic mechanism of the radiopharmaceutical, estimated radiation absorbed dose, visual description/interpretation, discussion, and references will be presented. In the atlas section, each image will have a legend describing the image, which will frequently be followed by a comment section. Although the introduction section may have a significant amount of text and information, the emphasis is on the images, with a significant portion of the chapter's text and information in the legend and comment section of each image. I believe this format will not only help the resident in learning a procedure or a specific aspect of a procedure in nuclear medicine, but the format will also help the experienced physician locate topics that are directly relevant to a particular clinical problem.

Finally, I welcome any comments regarding the series and volumes, and I solicit suggestions for future atlases.

Douglas Van Nostrand
Series Editor

Preface

In this initial volume of a series of selected atlases of nuclear medicine, five topics are presented, which should be valuable for the clinical practice of nuclear medicine.

Cholescintigraphy not only continues to remain a valuable diagnostic tool in the evaluation of acute cholecystitis, but it has also been found clinically useful in other disease entities. The first chapter, "Atlas of Cholescintigraphy: Selective Update," will discuss and illustrate not only recent advances in cholescintigraphy with emphasis on newly described scintigraphic findings and techniques but also new clinical indications for cholescintigraphy.

In the second chapter entitled "Atlas of Gastrointestinal Bleeding (RBC) Scintigraphy," Drs. Wiest and Hartshorne present the spectrum of scintigraphic findings and—most importantly—pitfalls in the interpretation of gastrointestinal bleeding scintigraphy.

The third chapter illustrates SPECT Tc-99m labeled red blood cell liver scintigraphy in the diagnosis of cavernous hemangiomas of the liver. In the clinical practice of nuclear medicine, SPECT Tc-99m labeled red blood cell liver scintigraphy has become an increasingly valuable and requested procedure. This atlas presents not only the spectrum of planar and SPECT scintigraphic findings but also potential problem areas and rare false positives.

The final two atlases describe procedures that one may occasionally be called upon to perform, and the atlases should offer an excellent source of information for review. The fourth chapter discusses and demonstrates Hepatic Arterial Perfusion Scintigraphy for evaluation of appropriate catheter placement and the adequacy of liver perfusion for intraarterial chemotherapy to the liver. The final chapter presents an atlas of Peritoneoscintigraphy. Although the latter has been used prior to an occasional intraperitoneal P-32 therapy, the frequency of peritoneoscintigraphy appears to be increasing in such areas as the evaluation of complications in patients on continuous ambulatory peritoneal dialysis.

Harvey Ziessman
Douglas Van Nostrand

Contents

Contributors

Jay Anderson, MD   Chief, Nuclear Medicine Service HSHL-XN, Walter Reed Army Medical Center, Washington, D.C. 20307; Associate Professor of Radiology and Nuclear Medicine, Uniformed Services University of Health Sciences, Bethesda, MD 20814, USA

Michael F. Hartshorne, MD   Chief, Radiology Service, Veterans Administration Medical Center, University of New Mexico, School of Medicine, Albuquerque, NM 87131, USA

Douglas Van Nostrand, MD   Director, Nuclear Medicine Department, Good Samaritan Hospital, Baltimore, MD 21239, USA; Clinical Professor of Radiology and Nuclear Medicine, Uniformed Services University of Health Sciences, Bethesda, MD 20814, USA

Philip W. Wiest, MD   Department of Radiology, University of New Mexico, School of Medicine, Albuquerque, NM 87131, USA

Harvey A. Ziessman, MD   Associate Professor of Radiology, Division of Nuclear Medicine, Georgetown University Hospital, Washington, D.C. 20007, USA

CHAPTER 1

Atlas of Cholescintigraphy: Selective Update

Harvey A. Ziessman

Cholescintigraphy has proven to be the imaging study of choice for the diagnosis of acute cholecystitis and has been found quite useful for the diagnosis and evaluation of a variety of other hepatobiliary diseases. Several atlases of radionuclide hepatobiliary imaging have previously been published.[1-3] This atlas describes and illustrates recent advances in cholescintigraphy. The emphasis is on newly described scintigraphic findings helpful in diagnosis as well as improvements or modifications in technique and computer processing.

Subjects to be covered will include the "rim" sign and radionuclide angiography in acute cholecystitis, morphine-augmented cholescintigraphy, the use of cholecystokinin to diagnose chronic acalculus cholecystitis, and the differential diagnosis of focal nodular hyperplasia, hepatic adenoma, and hepatocellular carcinoma with cholescintigraphy. Also included are examples of sclerosing cholangitis, postoperative biliary leaks, and causes for the postcholecystectomy syndrome (e.g., cystic duct remnants, partial common duct obstruction, and sphincter of Oddi dysfunction). Finally, quantitative cholescintigraphy, including deconvolution, is discussed and illustrated.

Technique

Radiopharmaceutical: Technetium-99m mebrofenin or Technetium-99m disofenin.
Administered dose:
Adults: Bilirubin
 <2.0 mg% 5.0 mCi (185 MBq)
 ≥ 2.0 mg% 7.5 mCi (278 MBq)
 ≥ 10 mg% 10.0 mCi (370 MBq).
Children: 200 μg/kg (no less than 1 mCi or 35 MBq).

Patient Preparation

1. Patients must fast for 4 hr prior to the study.
2. Patients who have not eaten within 24 hr prior to the study should receive sincalide (Kinevac, E. R. Squibb & Sons, Princeton, NJ) 0.02 μg/kg in 20 to 30 cc normal saline as a slow 3-min infusion. Radiotracer injection and imaging may begin 30 min after injection to allow time for gallbladder relaxation. Sincalide, the C-terminal octapeptide

1

of cholecystokinin (CCK) is the only commercially available CCK product in the United States.

Set-Up

Camera

1. 15% window over 140 keV photopeak
2. Large field-of-view gamma camera with an all-purpose parallel hole collimator

Computer

Two-phase study (1-sec frames for 1 min followed by 1-min frames for 59 min. For postcholecystectomy patients no flow study is performed.

Analog Images

Two-sec flow images for 1 min followed by an immmediate image for 500,000 counts, a 5-min image, and then images every 10 min for equal time.

Patient

Position the patient supine with inferior portion of the heart, entire liver, and abdomen in the camera field of view.

Procedure

1. Inject the radiopharmaceutical as an intravenous bolus and start the camera and computer.
2. Observe the persistence scope or computer monitor.
3. If the study is performed for suspected acute cholecystitis and the gallbladder has not visualized by 20 to 30 min (but there has been biliary to bowel transit), inject morphine sulfate intravenously, 0.04 mg/kg (up to 2 mg).
4. At the end of the 60 min, acquire a right lateral and left anterior oblique view.
5. If there is overlap of duodenal and common duct activity with gallbladder fossa making diagnosis difficult, give the patient water by mouth to clear the duodenal activity. Then reimage. Upright imaging may also be helpful.
6. Perform delayed imaging if:

    a. morphine was not administered and the gallbladder did not visualize. Reimage at 2 and up to 4 hr. Shielding of bowel activity and a longer acquisition time may be necessary to visualize the gallbladder fossa well if most of the tracer has cleared from the liver. Very rarely 24-hr images may be helpful.

    b. otherwise clinically indicated (e.g., hepatic insufficiency, partial common duct obstruction, patients on hyperalimentation, for detection of a biliary leak or cystic duct remnant, etc.).

CCK after Gallbladder Visualization

Indications:

1. This may be useful when the gallbladder and common duct have filled but there is no small bowel visualization. CCK can determine quickly whether or not there is a significant fixed obstruction at the sphincter of Oddi.
2. In the setting of acute acalculus cholecystitis, occasional false negative studies may occur as a result of incomplete functional obstruction. A noncontracting gallbladder would suggest cholecystitis.[4]
3. Calculation of gallbladder ejection fraction when chronic acalculus cholecystitis or the cystic duct syndrome is suspected.

Caution

CCK should not be administered if the patient has received morphine, since the prolonged effect of morphine on the sphincter of Oddi may give erroneous information on gallbladder and sphincter function.

Procedure for the Use of CCK

1. Set up the computer for an additional 20 to 30 1-min frames and the formatter for analog images every 5 min.
2. Start the computer 1 min before injection. Inject 0.02 μg/kg sincalide diluted in 20 to 30 cc slowly over 3 min. Rapid injection may cause spasm of the cystic duct.
3. Calculate the percent gallbladder emptying (first frame counts minus the last frame counts divided by the first frame counts corrected for background).

Image Interpretation

Review of the computer cinematic display can sometimes be very helpful diagnostically. Intermittent analog images may sometimes be unclear as a result of overlying small bowel activity. A cine playback can be particularly helpful in differentiating gallbladder filling from activity transiting the second portion of the duodenum and in evaluating the biliary flow pattern after biliary diversion operations.

Physiological Mechanism of the Radiopharmaceutical

Iminodiacetic acid (IDA) acts as a bifunctional chelate. At one end it attaches to ^{99m}Tc and at the other end it attaches to an analog of lidocaine carrying the biological function. IDA, an organic anion, is loosely bound to albumin in the bloodstream and is transported across the hepatocyte by the same carrier-mediated clearance mechanism as bilirubin. After extraction, the agent is transported across the hepatocyte and secreted into the bile canaliculi chemically unchanged. Hyperbilirubinemia competitively interferes with ^{99m}Tc IDA transport to varying degrees depending on the specific analogue used. Substitutions in the phenyl ring have wide impact on the uptake, retention, and excretion times for different IDA agents.[5]

Although many hepatobiliary radionuclides have been investigated, only two have been approved by the Food and Drug Administration (FDA) for clinical use in the United States (see Figure 1.1). The first was ^{99m}Tc disofenin or o-diisopropyl IDA (DISIDA) (Hepatolite, E.I. duPont). [99m]Tc DISIDA has been used successfully for a number of years. It has a high hepatic extraction and is useful even with bilirubin levels of 20 mg% and higher. Recently a second IDA derivative has been approved, ^{99m}Tc mebrofenin (m-bromo-o,p-trimethyl) IDA (Choletec, E.R. Squibb Diagnostics). [99m]Tc mebrofenin has somewhat higher hepatic extraction (98% vs. 82–88%), faster excretion (mean $t\frac{1}{2}$ of 17 min vs. 19 min) and less renal excretion (<1% in 3 hr vs. <9% in 2 hr) compared to DISIDA.[5,6] The difference between the two is not usually clinically important, except perhaps in patients with very poor hepatic function or when a low administered dose is used (e.g., in pediatric patients studied for biliary atresia). It is theoretically possible that ^{99m}Tc mebrofenin may be less useful for detecting delayed gallbladder visualization because of its more rapid hepatic clearance. With the use of morphine (discussed below), this should not be a problem.

The estimated radiation absorbed doses for Tc-99m disofenin and Tc-99m mebrofenin are shown in Table 1.

Table 1.1. Estimated radiation-absorbed doses.[a]

Organ	[99m]Tc disofenin (Hepatolite, E.I. duPont, Billerica, MA) rads/ 5 mCi	mGy/ 185 MBq	[99m]Tc mebrofenin (Choletec, Squibb Diagnostics, New Brunswick, NJ) rads/5 mCi	mGy/ 185 MBq
Total body	0.08	0.8	0.10	1.0
Liver	0.19	1.9	0.23	2.3
Gallbladder wall	0.60	6.0	0.63	6.3
Small intestine	1.06	10.6	1.50	15.0
Upper large intestine	1.90	19.0	2.37	23.7
Lower large intestine	1.38	13.8	1.82	18.2
Urinary bladder wall	0.46	4.6	0.15	1.5
Ovaries	0.41	4.1	0.51	5.1
Testes	0.03	3.0	0.03	0.3
Red marrow	0.14	1.4	0.15	1.5

[a]Estimates adapted from FDA-approved radiopharmaceutical package inserts.

Visual Description and Interpretation

New techniques and scintigraphic findings to help diagnose acute cholecystitis:

Morphine-Augmented Cholescintigraphy

Cholescintigraphy has proven very accurate for the diagnosis of acute cholecystitis. False negative studies are quite rare; however, false positives do occasionally occur.[7,8] Some false positives are unavoidable and one needs to be aware of these situations, e.g., prolonged fasting (>24 hr), hyperalimentation, severe hepatic insufficiency, alcoholism, pancreatitis (controversial), and acute intercurrent illness.[9] The administration of cholecystokinin (CCK) to patients who have been fasting for >24 hr in an attempt to empty the gallbladder may prevent some false positives; however, it should be kept in mind that this may not always be effective (e.g., in patients with chronic cholecystitis and a noncontracting gallbladder). One should also be aware that CCK administered prior to the study may delay biliary to bowel transit and give the appearance of partial common duct obstruction.[10] Other causes of false positive studies can be avoided if anticipated, e.g., by requiring the patient to be N.P.O. for at least 4 hr prior to the study and obtaining images for 2 to 4 hr after injection to detect delayed gallbladder visualization. Although less than 5% of patients with chronic cholecystitis will have delayed gallbladder visualization, emergency surgery for acute cholecystitis may be avoided in these cases.[9] The overall false positive rate can be decreased by 10% by obtaining 2- to 4-hr delayed views.[8,9] Occasionally gallbladder visualization will occur as late as 24 hr, especially in patients with hepatic insufficiency or intercurrent illness.[11] However, this diagnostic delay in acutely ill patients is not ideal from the viewpoint of the clinician and surgeon.

As an alternative to delayed imaging, Choy et al. have reported the interventional use of low-dose morphine, allowing completion of the study by 90 min or earlier[12] (see Figures 1.2, 1.3, 1.4). Since morphine increases the intraductal pressure at the sphincter of Oddi, biliary drainage will preferentially empty into the cystic duct if it is not obstructed. Typically the gallbladder will fill within 5 to 10 min after morphine injection. Although occasional false positives may still occur in the setting of hepatic insufficiency, severe intercurrent illness, or chronic cholecystitis,[13,14] many studies have now reported morphine cholescintigraphy to be as accurate as using the delayed imaging method.[15,16] There has been concern that morphine might increase the false negative rate (i.e., gallbladder filling with acute cholecystitis) by causing dislodgement of a stone from the cystic duct; however, this does not seem to be a common problem[17] (see Figure 1.5).

"Rim" Sign and Radionuclide Angiography

Two ancillary diagnostic findings have been described with acute cholecystitis, the "rim" sign and increased blood flow to the gallbladder fossa on radionuclide angiography (see Figure 1.6).

A rim of increased hepatic uptake and retention adjacent to the gallbladder fossa has been described as a useful secondary sign of acute cholecystitis.[18,19] It has been reported to be present in about one third of patients with nonvisualization of the gallbladder and acute cholecystitis.[20,21] This finding has been associated with an increased incidence of perforation and gangrene.[18,19] However, it may also be seen in severe acute cholecystitis without these complications.[20,21] This is a very specific scintigraphic finding for the diagnosis of acute cholecystitis and its presence can increase one's certainty about the diagnosis, particularly in a patient who may have an increased likelihood of a false positive study for reasons stated above. In addition to giving prognostic information regarding the seriousness of the acute process and the potential for complications, its high predictive value for acute cholecystitis may obviate the need for 2- to 4-hr delayed imaging.[22]

Colletti et al. described increased blood flow to the region of the gallbladder fossa in acute cholecystitis.[23,24] The presence of this finding increased the positive predictive value of nonvisualization of the gallbladder for acute cholecystitis from 71% to 85%. However, radionuclide angiography had a 20% false positive rate. Also, although 88% of patients with gangrene had increased flow, only 29% of patients with increased flow had a gangrenous gallbladder. Increased blood flow was present in all patients with a positive "rim" sign. Although we routinely perform flow studies, we have not seen this sign as frequently as reported. As Colletti et al. point out, one must be sure that the apparent increased flow is not the result of renal blood flow. When present in the setting of nonvisualization of the gallbladder, the finding can increase the certainty that this is indeed acute cholecystitis and can shorten the study when the delayed imaging technique is used.

Focal Nodular Hyperplasia, Hepatic Adenoma, Hepatoma: Differential Diagnosis with Cholescintigraphy

Focal nodular hyperplasia (FNH) is a benign tumor of the liver. In the past it was confused clinically and pathologically with hepatic adenoma. However, it is now known that they are distinct entities with very different prognoses. FNH, comprised of hepatocytes, Kupfer cells, and bile ducts, is usually asymptomatic and found incidentally. In contrast, hepatic adenoma, also a benign tumor, but comprised only of hepatocytes, has been associated with birth control pill use and life-threatening hemorrhage. ^{99m}Tc sulfur colloid (^{99m}Tc SC) can sometimes help differentiate these two entities since hepatic adenomas do not typically take up ^{99m}Tc SC (no Kupfer cells), whereas FNH takes up the radiopharmaceutical in approximately two thirds of cases and even occasionally has increased uptake.[25] Hepatomas, like all malignant tumors, are cold with ^{99m}Tc SC.

Cholescintigraphy may have an important role in the differential diagnosis of these tumors. On cholescintigraphy, tumors of FNH have increased blood flow and prompt normal extraction, but delayed biliary clearance[26,27] (see Figure 1.10). This is probably a result of normal hepatocyte function, but disordered biliary drainage. In contrast, hepatomas may have increased flow, but images over the first hour generally show no uptake. However, 2-hr delayed imaging frequently shows uptake while the adjacent normal liver has cleared of radiotracer[28,29] (see Figure 1.11). This pattern of delayed uptake and clearance may be due to the tumor's relatively poor function compared to normal liver. The cholescintigraphic findings with hepatic adenomas are less well defined. Only a few have been reported and well documented.[30-32] However, a large European study using cholescintigraphy to study hepatic tumors (72 with FNH, 5 hepatic adenomas, and 67 hepatomas) found a sensitivity of

83% and specificity of 100% for diagnosing FNH, using the above criteria. The hepatic adenomas showed either decreased or no uptake on cholescintigraphy.[31,32] This pattern of hepatic adenoma overlaps with that of hepatoma but not FNH. Therefore, FNH usually can be differentiated and unnecessary surgery avoided.

Sclerosing Cholangitis

Sclerosing cholangitis is an uncommon disease of unknown etiology characterized by progressive inflammatory fibrosis of the biliary tree leading to diffuse stricture formation. Symptoms include vague abdominal discomfort, fatigue, pruritus, and jaundice. Characteristically, these patients have prominent elevation of alkaline phosphatase, while bilirubin rises late in the disease. The clinical findings are not specific and overlap with isolated common bile stricture and primary biliary cirrhosis. Definitive diagnosis requires cholangiography with demonstration of diffuse multifocal strictures of the biliary tree. However, this provides only morphological information, not physiological information about hepatic bile flow or gallbladder function. Full assessment of the intrahepatic biliary tree is often limited by dominant tight strictures at the common or proximal right and left hepatic ducts that prevent adequate filling of ducts with contrast.

Recently, cholescintigraphy has been shown to be a sensitive test for diagnosing sclerosing cholangitis, providing morphological and physiological information regarding the extent and severity of disease[33] (see Figure 1.12). Cholescintigraphy is able to distinguish sclerosing cholangitis from isolated common bile duct obstruction and primary biliary cirrhosis, allowing selection of patients for cholangiography or permitting noninvasive diagnosis in patients unwilling or unable to undergo cholangiography. Cholescintigraphy is also useful for following a patient's clinical course and it allows for the noninvasive evaluation of the effectiveness of therapeutic interventions (e.g., balloon dilatation of localized stenoses).

The pattern of sclerosing cholangitis on cholescintigraphy is that of diffuse and focal areas of segmental obstruction with delayed hepatobiliary clearance. These regional abnormalities can be quantified and used as indicators of disease progression and may indicate the need for intervention and be used to evaluate the effectiveness of therapy.

Chronic Acalculus Cholecystitis: Diagnosis with CCK-Stimulated Cholescintigraphy and Calculation of a Gallbladder Ejection Fraction

A difficult and not uncommon clinical problem is the patient with symptoms of chronic recurrent right upper quadrant pain suggestive of biliary colic. However, repeated extensive workups, including ultrasonography, cholescintigraphy, and even oral cholecystography reveal no abnormality. Cholecystectomy frequently cures the pain. Pathologically, one finds hypertrophy of the gallbladder wall and muscularis propria, with or without monocellular infiltration, serosal thickening, Aschoff-Rokitansky sinuses, foamy macrophages filling the tips of mucosal folds, and yellow papillary nodules (cholesterolosis). However, no gallstones are present. The pathological diagnosis is chronic acalculus cholecystitis.[34] This diagnostic dilemma can now be noninvasively resolved preoperatively by doing a simple functional test of gallbladder function, that is, CCK-stimulated cholescintigraphy with calculation of a gallbladder ejection fraction (see Figures 1.7 and 1.9). A diseased, symptomatic gallbladder with chronic acalculus cholecystitis contracts suboptimally, with an ejection fraction usually <35%. Several studies have reported excellent results in predicting acalculus biliary disease preoperatively using these criteria.[35-38] It should be pointed out, however, that there are some other less enthusiastic reports.[39,40]

Symptoms of nausea, vomiting, and abdominal pain mimicking the patient's recurrent symptoms may occur during the 3-min administration of CCK. One should be cautious about assuming the diagnostic value of these symptoms since CCK causes not only gallbladder contraction and sphincter of Oddi relaxation, but it also increases peristalsis of the small and large bowel and stimulates pancreatic secretion. Therefore, the elicited pain is probably nonspecific.

It is interesting to note that about half of patients with gallstones will also have a reduced ejection fraction. This technique might also be a way to select patients with symptomatic chronic calculus cholecystitis who could benefit from elective cholecystectomy from those with abdominal pain of various etiologies but only incidental gallstones. However, this is only speculative.

Postoperative Biliary Leaks

Early detection of postoperative leaks can significantly reduce patient morbidity and mortality. Cholescintigraphy is an accurate method to detect the presence of a bile leak and determine the site of leakage[41,42] (see Figures 1.13 and 1.14). Unlike other morphological imaging modalities, cholescintigraphy can confirm that a fluid collection is indeed of biliary origin. Intraabdominal bile leaks may not be clinically significant if sterile and if the amount of leakage is small. They frequently seal spontaneously. However, with a large amount of extravasation and little bile entering the intestine, surgical intervention is frequently necessary. Cholescintigraphy can be helpful in making this differentiation.[41-43] Delayed views are often needed.

Postcholecystectomy Pain Syndrome

Postcholecystectomy pain syndrome is another clinical condition that is frequently frustrating to both patient and physician. Recurrent pain after cholecystectomy has numerous etiologies, including a diseased cystic duct remnant, retained or recurrent common duct stones, inflammatory stricture of the common duct, bile gastritis, and sphincter of Oddi dysfunction.

Cystic Duct Remnant

A postoperative cystic duct remnant can act as a small gallbladder and become diseased, producing symptoms identical to those of acute and chronic cholecystitis. Cystic duct remnants are not uncommon, occurring in one series in 14% of 125 symptomatic postcholecystectomy patients.[41] Cholescintigraphy is useful for detecting a symptomatic remnant[44] (see Figure 1.15). However, it is essential to obtain delayed views for as long as activity remains in the hepatobiliary system to ensure visualization of the remnant.

Partial Common Duct Obstruction

A retained or recurrent common duct stone or inflammatory fibrosis is a common cause of recurrent postcholecystectomy pain. Cholescintigraphy has been shown quite useful as a noninvasive method for diagnosing partial biliary obstruction (see Figure 1.16). Complete obstruction is typically not difficult to diagnose. Ultrasonography usually will demonstrate a dilated common bile duct and cholescintigraphy will show good uptake but no clearance. However, a partial common duct obstruction can be a more difficult problem. Ultrasound may or may not show dilated ducts; common duct stones are infrequently seen. Cholescintigraphy has an important role to play here. Typical findings of partial common bile duct obstruction may include bile pooling in the ducts proximal to the site of obstruction, an apparent cutoff at the site of obstruction, abnormal ductal dynamics with increasing rather than decreasing activity between 1 and 2 hr after injection, and delayed biliary to bowel transit.[41,45-47] Recent reports have found quantitative cholescintigraphy helpful in diagnosing partial common duct obstruction.[47-49] Although methods of quantitation vary, typically regions of interest are drawn on computer for the liver, common duct, and combined liver and common duct. Time–activity curves are generated and various quantitative parameters calculated, for example, time to peak, time of half emptying, and percent emptying at 60 min. Since the reported methods have varied considerably, normal controls are needed.

Sphincter of Oddi Dysfunction

Sphincter of Oddi dysfunction is felt by some to be a common cause of the postcholecystectomy syndrome. It is estimated to underlie as many as 13.8% of those with postcholecystectomy pain.[50] Because this is an imprecisely defined clinical syndrome, a variety of terms have been used to describe it, including papillary stenosis, biliary spasm, and biliary dyskinesia. Clinically, patients experience recurrent attacks of biliary-type pain after cholecystectomy but have no demonstrable mechanical cause.

The diagnosis of sphincter of Oddi dysfunction is difficult. It has been associated with a dilated common duct (nonspecific), a delay in emptying of x-ray contrast (>45 min) after cannulation of the sphincter during endoscopic retrograde cholangiopancreatography (ERCP) (hard to quantify), and elevated basal sphincter of Oddi manometry pressure (>40 mm Hg). A variety of sphincter abnormalities have been described: a) papillary stenosis, a fixed partial obstruction, b) dyskinesia, if the elevated basal pressure decreases in response to a sphincter relaxant (e.g., nitrates or CCK), and c) a paradoxical response to CCK, that is, an increase, rather than the normal decrease, in basal sphincter pressure after CCK injection. The latter, although frequently discussed, is not well documented, since the studies reporting this finding gave CCK as a rapid bolus (over seconds), a method known to cause common duct spasm in normal individuals.[51,52] Others have not confirmed this paradoxical response to CCK.[53]

ERCP with sphincter of Oddi manometry is considered by some to be the gold standard for diagnosing sphincter dysfunction; however, the procedure is difficult to perform, has a failure rate of about 20%, is poorly standardized, and occasionally induces pancreatitis. Few normal subjects or patients with other clearly defined recurrent pain syndromes have been studied; therefore the specificity of elevated sphincter pressure is not certain.[46,50,51]

Cholescintigraphy has been reported to be quite useful for diagnosing sphincter of Oddi dysfunction[46,53-56] (see Figures 1.17 and 1.18). However, since one would expect partial common bile duct obstructions, regardless of the etiology, to have a similar pattern, it is necessary to rule out a mechanical cause for obstruction. Quantitative cholescintigraphy has been found helpful in selected groups of patients to help diagnose sphincter of Oddi dysfunction.[53,55] Quantitation may be particularly helpful in choosing patients for invasive ERCP, confirming ERCP results, and in the follow-up of these patients postsphincterotomy.[53,57] CCK-stimulated quantitative cholescintigraphy may play a role in differentiating a fixed obstruction from a dyskinetic sphincter. This might be particularly important if pharmacological therapeutic trials are instituted.[50] At present, long-term response to treatment is probably the only true gold standard for final diagnosis of this perplexing syndrome.

Sphincterotomy, operative or endoscopic, is the treatment of choice, although proof of its efficacy is somewhat mixed.[57-59] A recent prospective double-blind study on the utility of sphincterotomy found a 91% clinical response rate in patients with elevated basal sphincter pressure.[58]

Enterogastric Biliary Reflux

Symptoms of gastritis may result from enterogastric reflux of bile. This is seen most commonly after gastric resection (e.g., Bilroth II surgery). Bile reflux has also been associated with gastritis symptoms after duodenal ulcer surgery and after cholecystectomy.[60] However, it may also occur in patients who have not had gastric surgery (e.g., it has been associated with acute cholecystitis).[61] Symptoms overlap with a variety of other gastrointestinal entities. Cholescintigraphy can be quite helpful in making this diagnosis (see Figure 1.19). Some bile reflux is commonly noted on cholescintigraphy, especially after the administration of CCK. This may not be clinically important; however, large volume reflux that clears poorly may be more important. Quantitation may be helpful.[62] We have seen delayed gastric emptying in a number of patients with bile reflux. Prolonged retention of bile in the stomach may play a role in symptomatic disease and may be a cause of symptoms postcholecystectomy.

Deconvolution Analysis

Deconvolutional analysis is a quantitative technique to help distinguish hepatocyte versus biliary disease (see Figure 1.20). The differentiation of primary biliary from primary hepatocyte disease with cholescintigraphy is not always easy. Primary biliary disease is characterized by relatively good hepatic extraction from the blood, but retention of radiotracer in the bile above the level of obstruction. In contrast, hepatocyte disease has impaired hepatic extraction with delayed clearance from the blood pool. But there may be considerable overlap, depending on the stage of disease. Deconvolution appears to increase the sensitivity for separating primary hepatocellular disease from primary biliary disease, especially in early biliary disease.[63]

Mathematically, the observed liver radioactivity at any time is a convolution of the true liver counts with the blood counts for all preceding time. Deconvolution attempts to derive the true liver counts from the observed liver counts, when the observed liver counts are masked by recirculation of the radiotracer within the liver pool. In other words, the deconvolved liver curve is the time activity curve that would be obtained with a perfect, infinitesimally short, bolus injection directly into the hepatic artery, with no recirculation of tracer. Once the deconvolutional analysis is completed, it is necessary to derive some quantitative parameters that describe the liver response curve. Brown et al. have defined the hepatic extraction fraction (HEF). HEF equals the ratio of the Y intercept of the exponential fit divided by the maximum Y data value of the liver response curve. Using the HEF and the liver excretion $t_{\frac{1}{2}}$, with image patterns, they were able to differentiate normals, cirrhotics, and patients with biliary obstruction.[63]

Acknowledgments. The author wishes to thank the following persons for their contributions to this atlas: Robert M. Albright, M.D., Stanley M. Knoll, M.D., Aaron M. Kistler, M.D., and Jack E. Juni, M.D. Special thanks go to Educational Media Production and Services at Georgetown University Medical Center for making this atlas possible.

Atlas Section

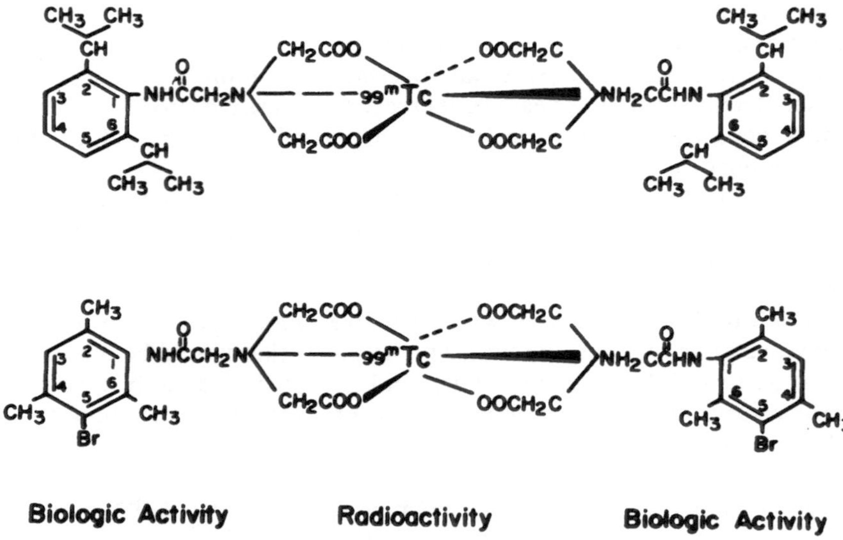

Figure 1.1. Structural configuration of ^{99m}Tc-labeled disofenin (*top*) and mebrofenin (*bottom*).

NCH$_2$COO (iminodiacetate) acts as a bifunctional chelate, carrying ^{99m}Tc (radioactivity) at one end and an acetanilide (biological activity) at the other end. Disofenin has CH-(CH$_3$)$_2$ at the 2 and 6 position, whereas mebrofenin has CH$_3$ at the 2, 4, and 6 positions and bromine at the 5 position. These minor structural differences between the two account for their major biological differences.

(Reprinted from Krishnamurthy S, Krishnamurthy GT. Quantitative assessment of hepatobiliary diseases with Tc-99m-IDA scintigraphy. In: Freeman LM, Weissman HS, eds. *Nuclear Medicine Annual* 1988. New York: Raven Press, p. 311. With permission.)

Biologic Activity   Radioactivity   Biologic Activity

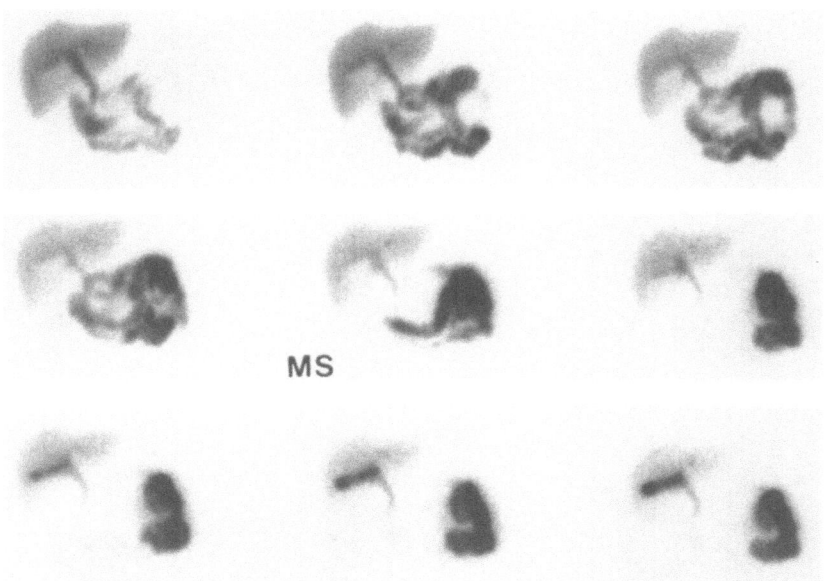

Figure 1.2. Morphine-augmented cholescintigraphy: negative.

Although there is prompt biliary clearance into the common duct and small bowel, the gallbladder was not seen by 40 min after ^{99m}Tc mebrofenin injection. Therefore, 2 mg morphine sulfate was administered intravenously (MS). Within 5 min of injection (*next image*), gallbladder filling begins. This is a true negative study for acute cholecystitis.

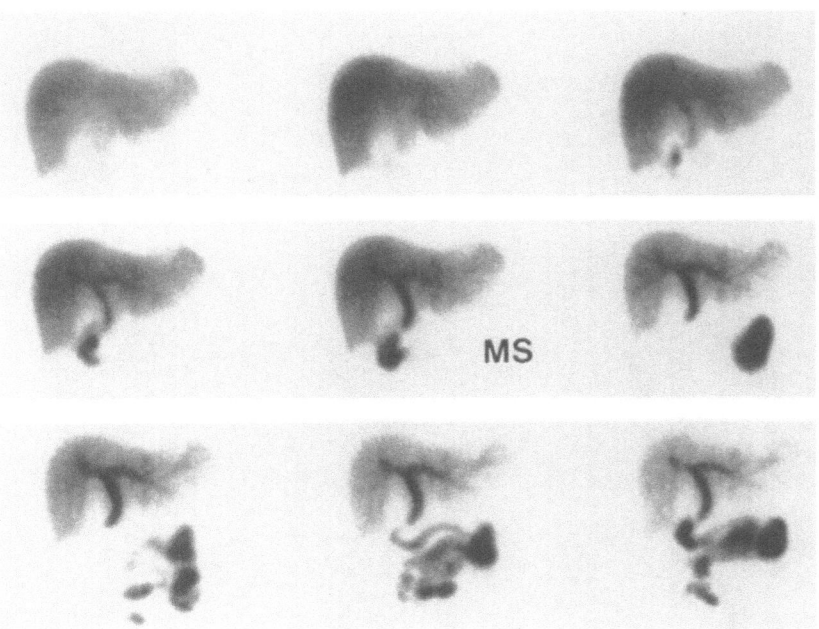

Figure 1.3. Morphine-augmented cholescintigraphy: positive.

Because of nonvisualization of the gallbladder, morphine (MS) was given at 30 min. The gallbladder does not fill for the remaining 30 min of the study. Note how the duodenum clears after morphine, an effect of morphine that can help in ambiguous cases. This is a true positive for acute cholecystitis.

Figure 1.4. Two sequential studies in the same patient, one without morphine (false positive) and then one with morphine (true negative for acute cholecystitis).

A: There is no gallbladder visualization at the end of this 60-min study and near complete washout of the tracer from the liver. Duodenal activity near the gallbladder fossa clears spontaneously. At this point, the study is positive for acute cholecystitis. Delayed imaging is indicated, but near complete washout makes that impossible.
B: The patient was reinjected with ^{99m}Tc mebrofenin and given morphine sulfate (MS). Prompt gallbladder visualization is seen (*arrowhead*). The morphine-augmented study is a true negative for acute cholecystitis.

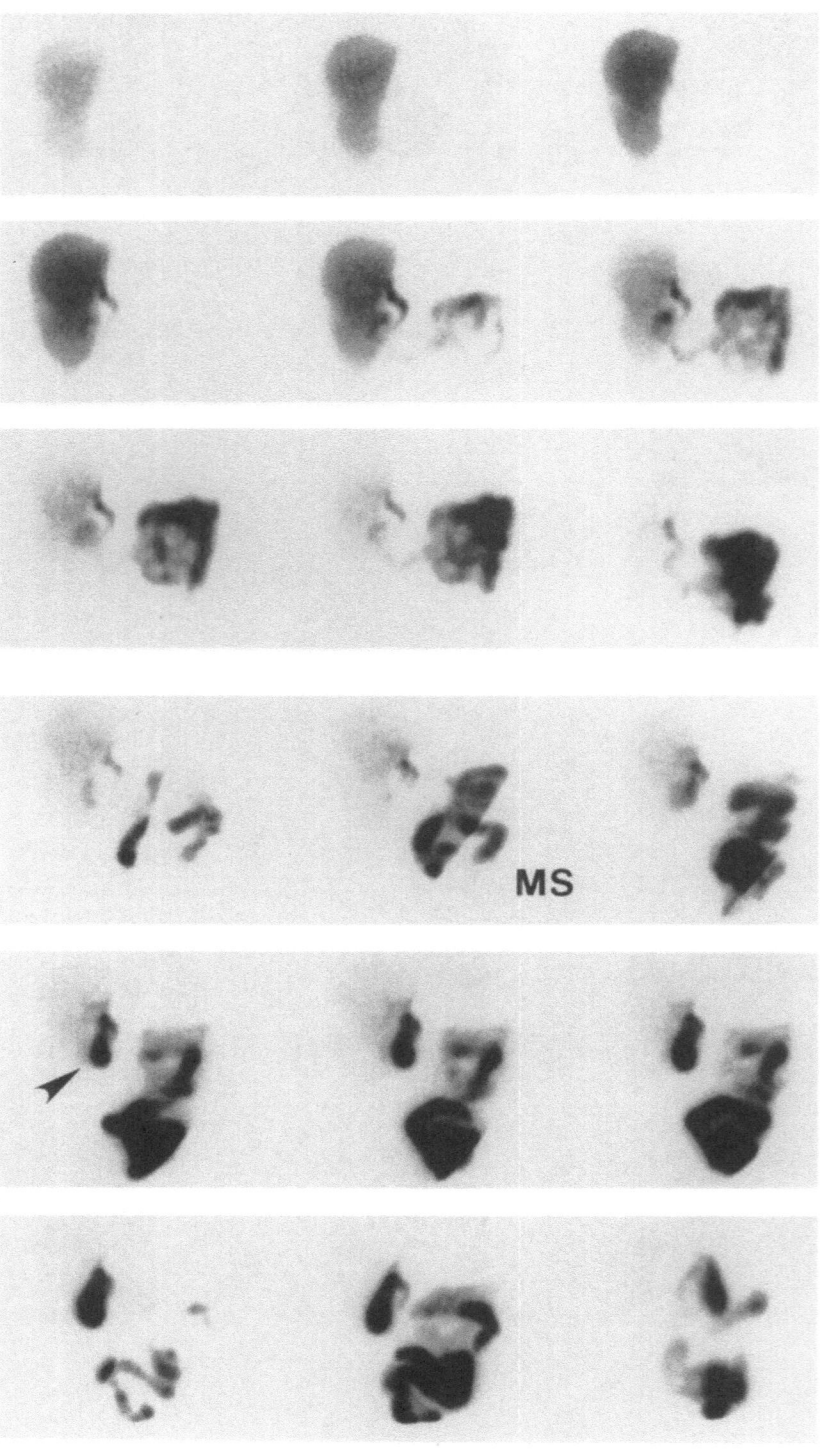

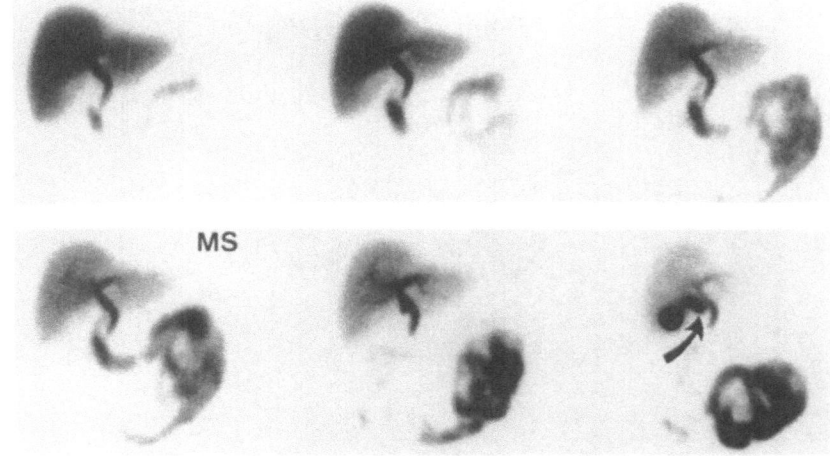

Figure 1.5. False negative HIDA.

Morphine (MS) was administered at 40 min. By 50 min the cystic duct is seen and the gallbladder clearly visualizes by 60 min after injection. However, note that the cystic duct appears dilated (*arrow*). This suggests partial or prior obstruction. Because of continued symptoms the patient went to surgery and had a cholecystectomy. Pathological exam diagnosed acute cholecystitis. It is possible that morphine dislodged the cystic duct stone.

(Reprinted from Kistler AM, Ziessman HA, Gooch D, Bitterman P. Morphine-augmented cholescintigraphy in acute cholecystitis: a satisfactory alternative to delayed imaging. *Clin Nucl Med.* 1991;16: 404–406. With permission.)

Figure 1.6. Acute cholecystitis with positive radionuclide angiogram and "rim" sign.

A: Two second/frame blood flow images show increased flow to the region of the gallbladder fossa (*arrows*).
B: Dynamic images every 5 min for 45 min after ^{99m}Tc disofenin injection show no definite visualization of the gallbladder, but there is increased activity around the gallbladder fossa (*arrow*).
C: Image at 60 min shows good hepatobiliary washout, but persistence of diffuse activity in the region of the gallbladder fossa, the "rim" sign (*arrow*). Acute cholecystitis was confirmed at surgery without evidence of gangrene or perforation.

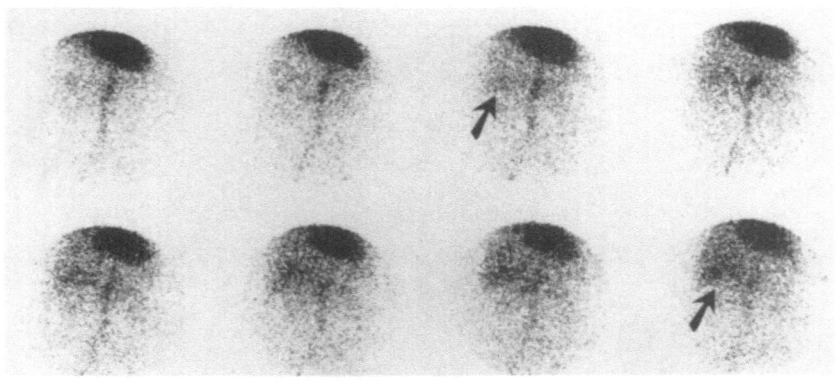

A

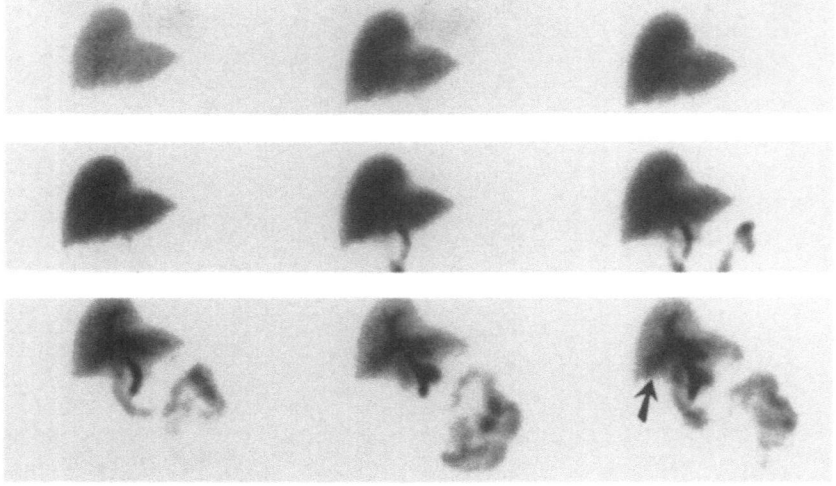

B

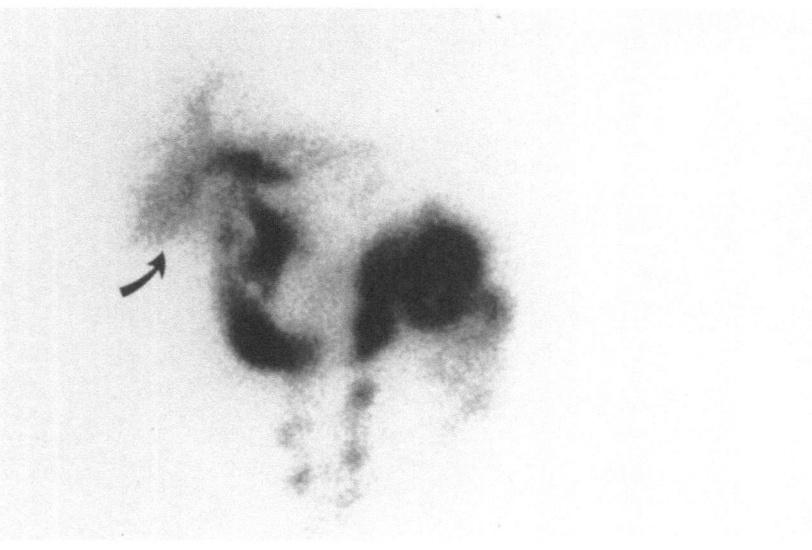

C

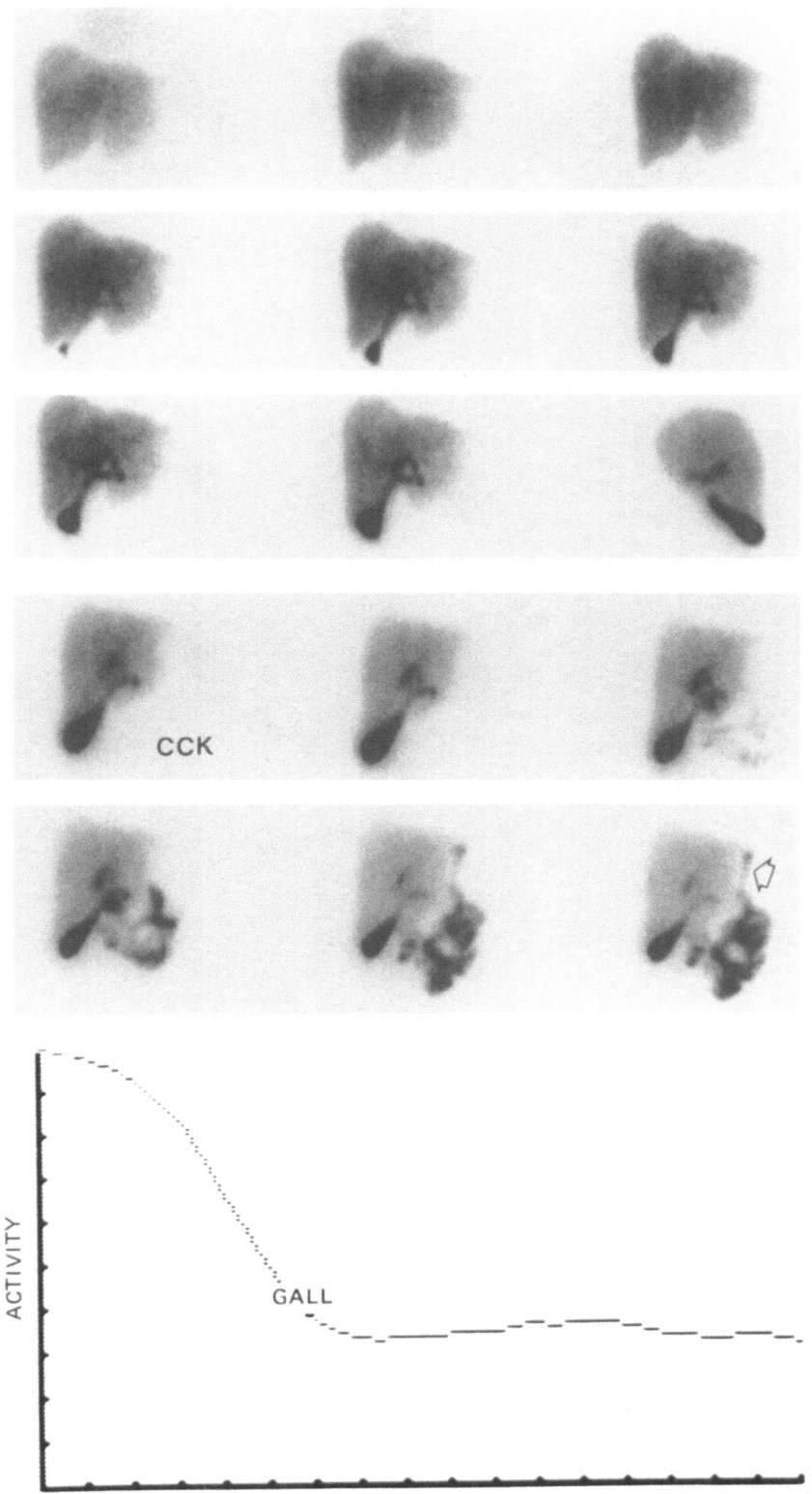

Figure 1.7. Normal variation versus common duct obstruction. Use of CCK.

A: Normal gallbladder filling at 60 min but delayed biliary to bowel transit. CCK was given to determine if there was common duct obstruction or merely a "hypertonic" sphincter and to evaluate gallbladder function.

B: Analog (LAO) images every 5 min acquired after a 3-min intravenous injection of 0.02 μg/kg sincalide (CCK). There is now good biliary to bowel transit with relaxation of the sphincter of Oddi. The gallbladder contracts normally (see **C**). Note the small amount of bile reflux (*open arrowhead*), seen commonly after CCK injection.

C: A region of interest was drawn around the gallbladder on computer and background corrected; this time-activity curve resulted. Emptying is maximal by 10 min after CCK injection. The ejection fraction was calculated to be 66%. The entire study was interpreted as normal.

Figure 1.8. Use of the LAO view for quantification of a gallbladder ejection fraction.

In the 60-min anterior view there is overlap of the gallbladder, common duct, and duodenum. The LAO view separates these structures and makes accurate calculation possible.

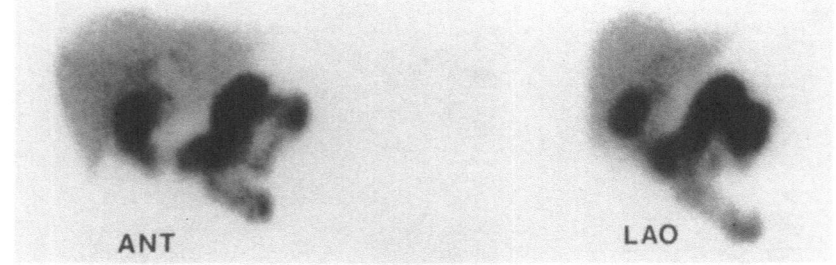

Figure 1.9. Chronic acalculus cholecystitis: abnormal ejection fraction.

This patient had chronic recurrent right upper quadrant pain with normal ultrasonography. The 60-min study (not shown) was normal with visualization of the gallbladder and normal biliary-to-bowel transit.
A: Sequential images every 5 min for 25 min after injection of CCK show no contraction of the gallbladder.
B: Gallbladder time–activity curve from the patient in A. Abnormal gallbladder ejection fraction is 16% (abnormal <35%). Surgery diagnosed chronic acalculus cholecystitis.

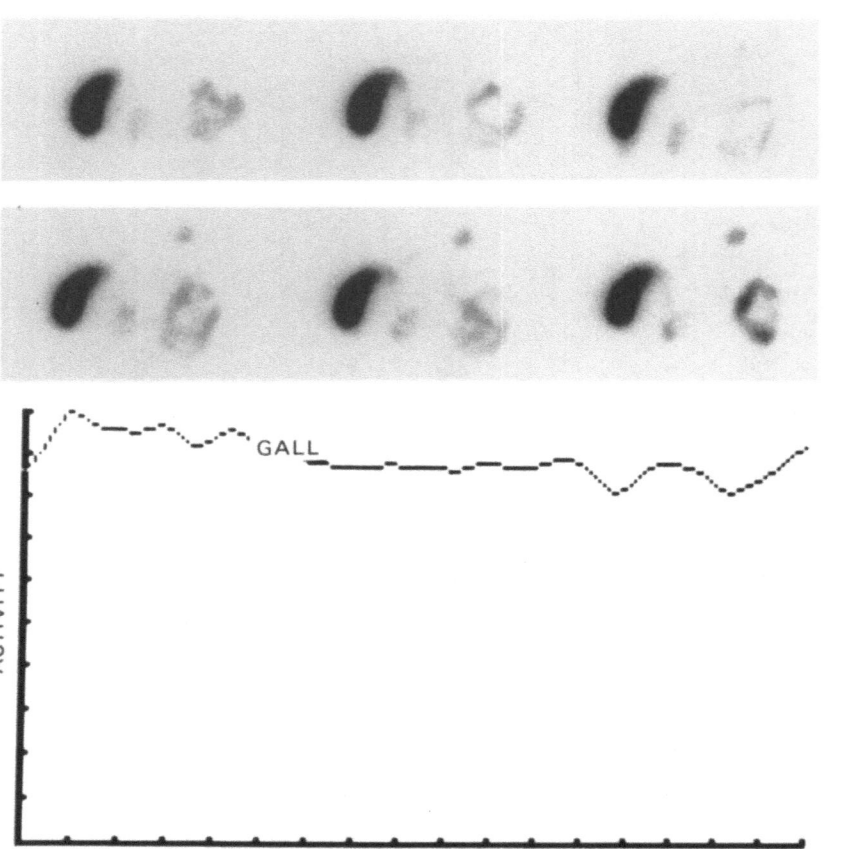

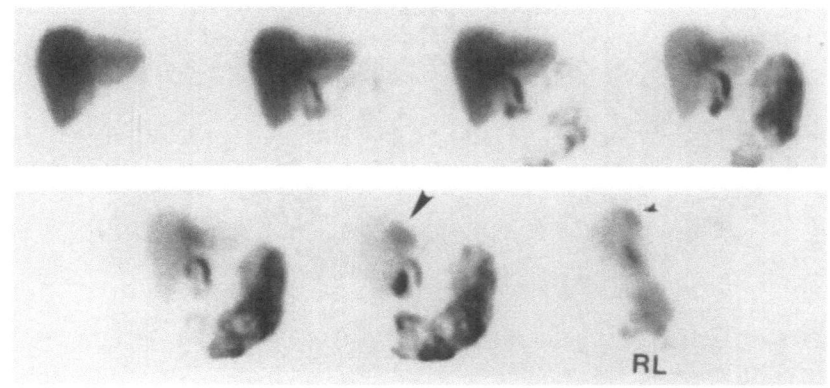

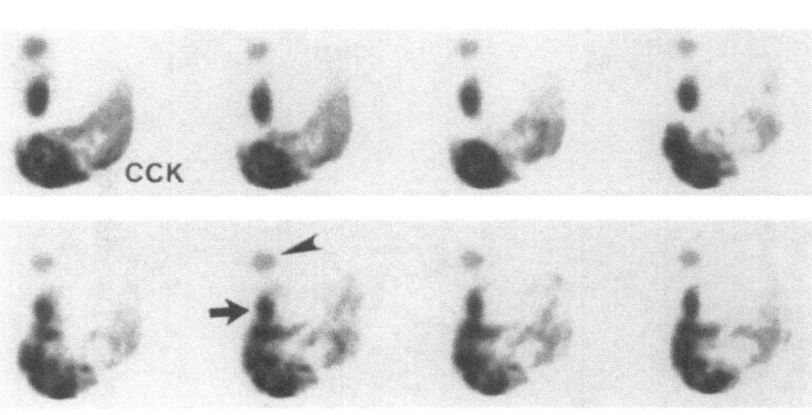

Figure 1.10. Focal nodular hyperplasia.

This patient was referred for symptoms of recurrent abdominal pain, a negative initial workup, and suspected chronic acalculus cholecystitis.

A: Images over 1 hr showed normal uptake, biliary clearance, and gallbladder filling. However, a focal area of delayed clearance is noted in the mid- to upper liver (*arrowhead*). The right lateral view shows this lesion to be anterior.

B: After injection of CCK to evaluate gallbladder contraction, the gallbladder (*arrow*) appears to contract normally (EF 45%) and most of the liver clears of tracer. However, the focus of delayed clearance persists (*arrowhead*).

Comment: This is the classic pattern for focal nodular hyperplasia. CT was negative. MRI showed a nonspecific mass in the left lobe. Since this patient had continued pain, which is not characteristic of focal nodular hyperplasia, the lesion was surgically removed. Focal nodular hyperplasia was diagnosed by pathologic exam.

Figure 1.11. Hepatocellular carcinoma (hepatoma).

A: Anterior view. A cold lesion (no uptake) is seen in the inferior lateral aspect of the right lobe (*curved arrow*) during the first 60 min of the study. At 2 hr, the hepatoma (*curved arrow*) has uptake while the remainder of the liver has cleared of tracer. The common duct (*open arrowhead*) and duodenum (*white arrow*) are noted.

B: Right anterior oblique view. The 60-min RAO view also shows a vague defect in the inferior lateral aspect of the right lobe (*arrow*). Imaging at 2 hr in the same view shows definite uptake within the tumor (*arrow*).

C: The CT scan shows the hepatoma (*arrowhead*) in the inferior aspect of the right lobe. The gallbladder is seen anterior medial to the tumor.

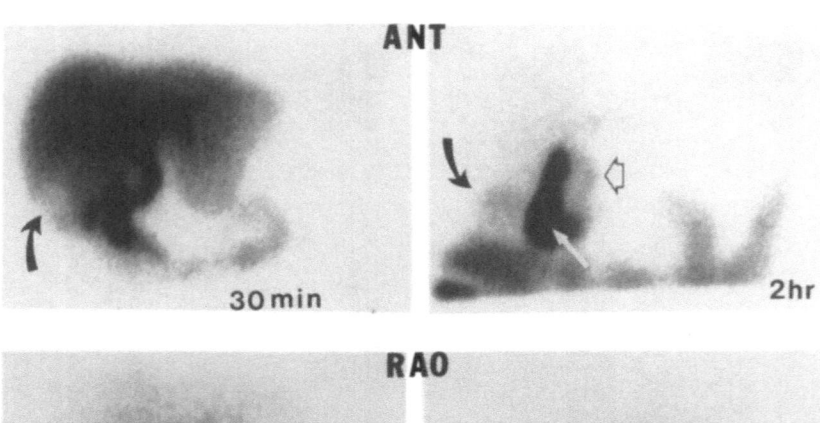

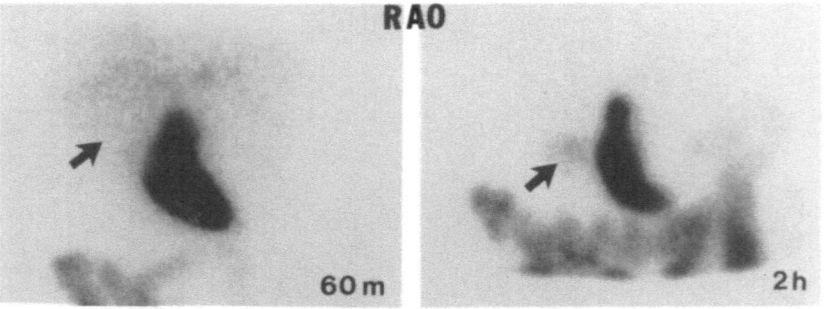

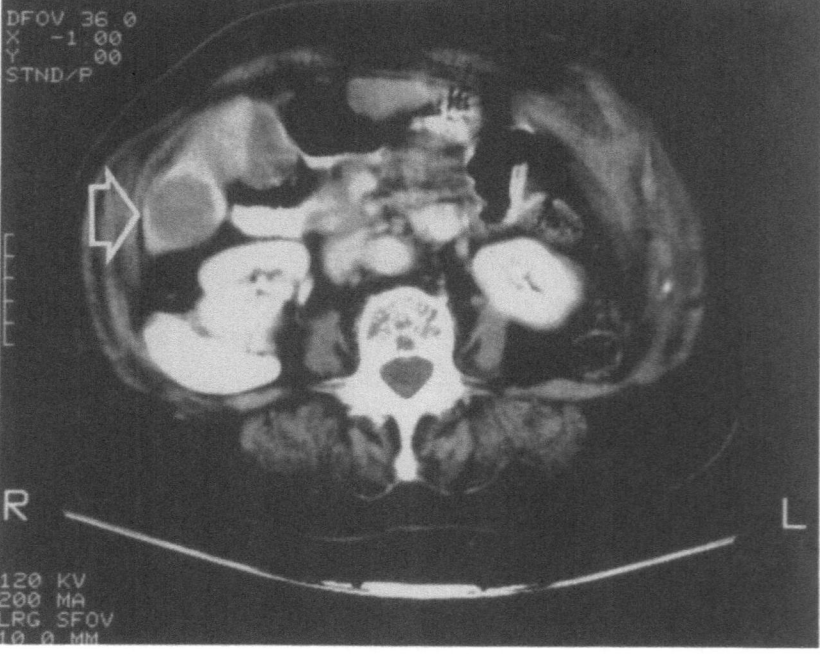

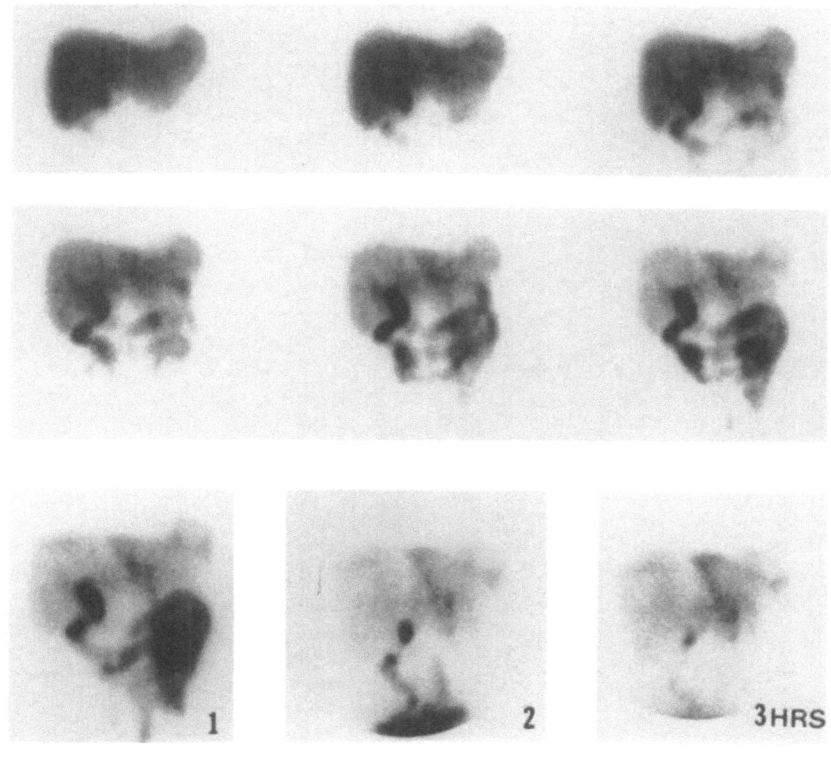

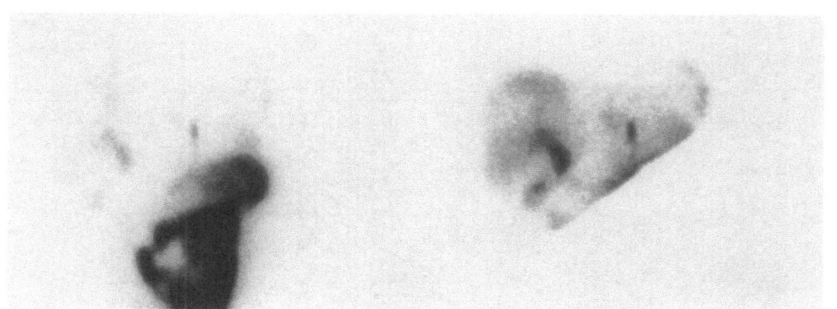

Figure 1.12. Sclerosing cholangitis.

Endoscopic retrograde cholangiopancreatography (ERCP) detected strictures in the upper right lobe, left lobe, and distal common duct. A biliary stent was placed in the common duct.

A: Sequential images every 10 min for 60 min show prominent intrahepatic ducts and heterogeneity of biliary clearance that appears particularly slow in the left lobe. The common duct appears dilated but has normal biliary-to-bowel transit.

B: Imaging at 1, 2, and 3 hr shows delayed heterogeneous clearance from the left lobe and to a lesser extent from the right. However, the common duct is clearing, showing the effectiveness of the biliary stent.

C: This 2-hr delayed image shows the utility of lead shielding and/or increasing the acquisition time when significant hepatobiliary washout has occurred. With no shielding (*left*), the liver and biliary ducts appear to be almost completely washed out. With proper shielding (*right*), it becomes obvious that there is still tracer in the liver, that clearance from both lobes is inhomogeneous, and that the common duct remains prominent. Also note the segmental biliary dilatation in the left lobe.

Figure 1.13. Postcholecystectomy biliary leak.

This elderly patient had a recent cholecystectomy and removal of a stone from the common bile duct. On postoperative day 8 he developed abdominal pain, tachypnea, tachycardia, and diaphoresis.

A: Serial images over 60 min show the site of active leak with increasing accumulation of radiotracer in the region of the cystic duct.

B: Delayed views at 2 hr [*Left:* anterior view with left side down. *Middle:* right lateral. *Right:* anterior (supine)]. These delayed views demonstrate increasing accumulation in two areas, the region of the cystic duct and gallbladder fossa anteriorly (*arrowheads*), as well as superiorly near the dome of the liver anteriorly and laterally (*arrows*), depending on the patient's position.

Comment: The patient was surgically explored, the cystic duct ligated, and drains placed. She recovered fully. The early images (**A**) delineate the site of active leakage. The delayed images with various views and different positions (**B**) demonstrate the extent of perihepatic and intraabdominal bile distribution.

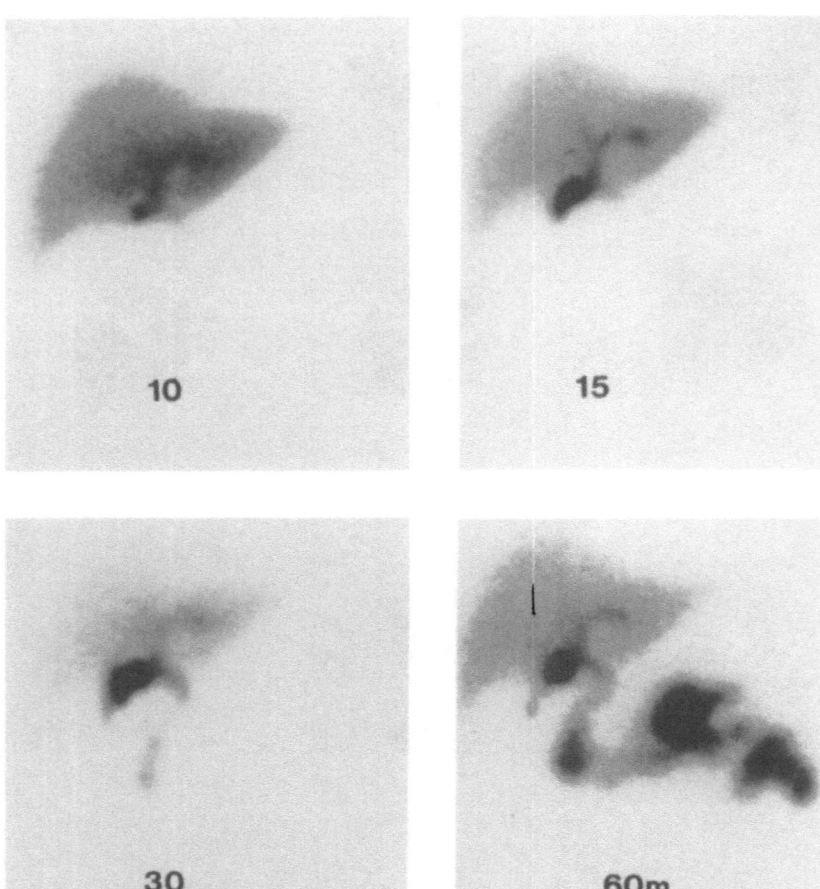

A

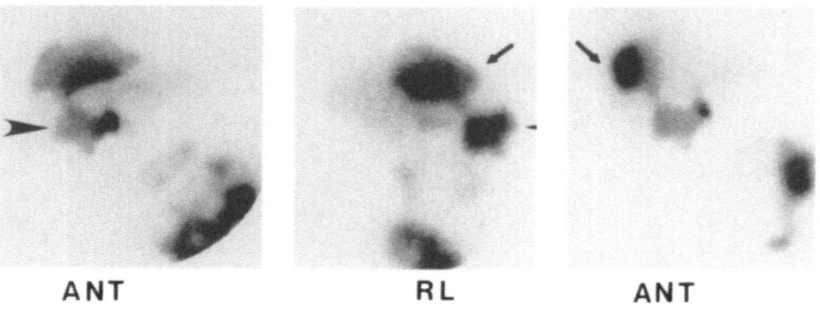

ANT   RL   ANT   B

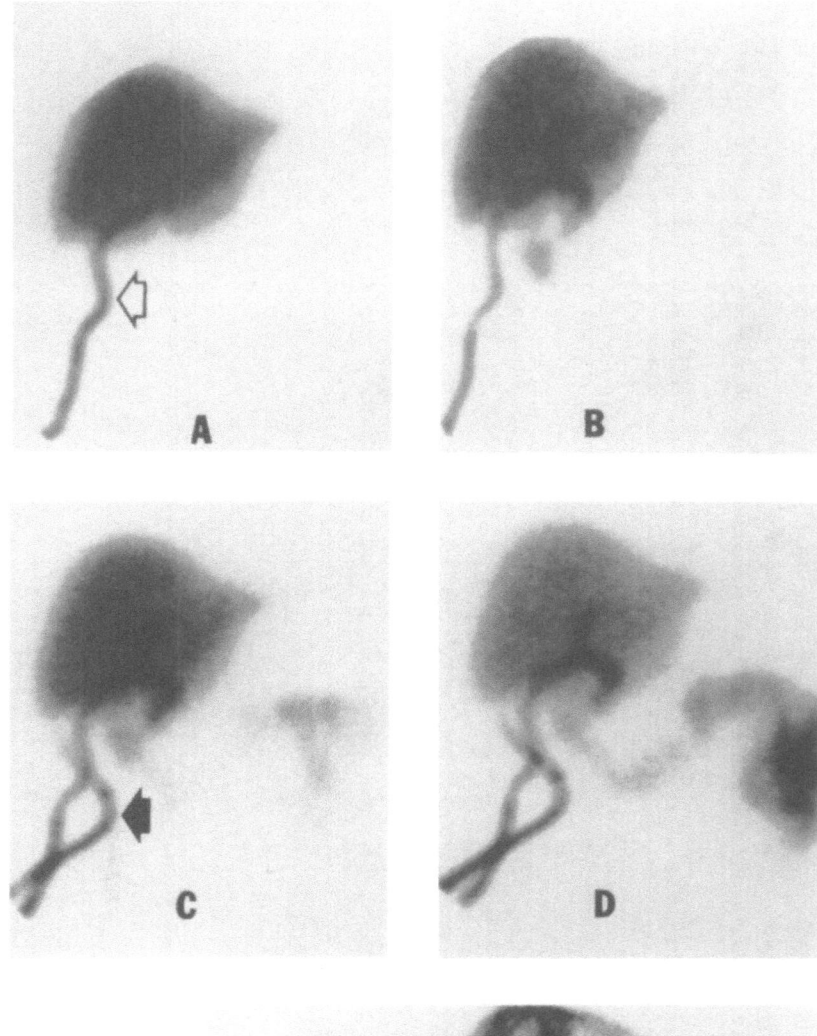

A

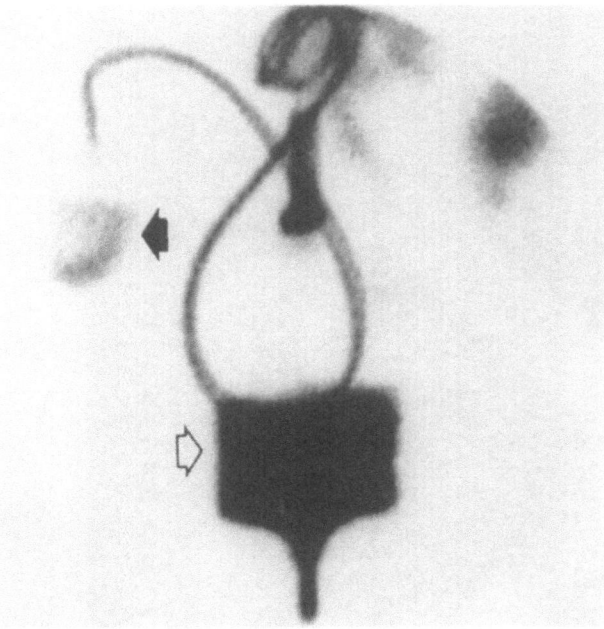

B

Figure 1.14. Postoperative cholecystectomy biliary leak.

A 64-year-old man had a cholecystectomy and common bile duct exploration with removal of two stones. On postoperative day 8 he developed fever and abdominal pain. Reexploration revealed a dislodged T-tube and bile peritonitis. The T-tube was repositioned and a Jackson–Pratt drain left in place. He continued to have fever, chills, nausea, diaphoresis, and abnormal liver function tests. Clinically viral hepatitis or hepatotoxicity secondary to cardiac drugs was suspected as the etiology. Cholescintigraphy was performed to rule out a biliary leak.

A: On the images performed 10 min after injection (**A**), there is only clearance into the patient's T-tube (*open arrowhead*). The T-tube was then occluded, followed by visualization of the common duct and prompt biliary-to-bowel transit on images at 20, 30, and 50 min (**B,C,D**). The second tube to visualize was the Jackson–Pratt drain (*closed arrowhead*).

B: Tubing and collection bags for T-tube (*closed arrowhead*) and Jackson–Pratt drainage (*open arrowhead*) both show radiotracer in them. A T-tube contrast study showed no obstruction of the common bile duct, but there was communication of the Jackson–Pratt drain with the common bile duct. With conservative medical management the patient improved.

Comment: This study emphasizes several important points: a) The nuclear medicine physician needs to know what the clinical question is, for example, in order to determine if there is bile flow through the common duct the T-tube must be closed, b) no intraabdominal collection may be detected if there is good drainage. Careful attention to the tubing and collection bags makes it possible to diagnose biliary leakage and even localize it by knowing the exact surgical placement of the drains.

Figure 1.15. Cystic duct remnant.

A: A 40-year-old man 15 years postcholecystectomy had symptoms of intermittent right upper quadrant pain and fatty food intolerance. Ultrasound was technically suboptimal due to the patient's obesity and excessive bowel gas. The common bile duct measured 9 mm. The ^{99m}Tc DISIDA study demonstrates normal heptatic uptake, clearance, and biliary-to-bowel transit time. However, a persistent focus of activity is identified anterolateral to the midportion of the common bile duct consistent with a cystic duct remnant (*arrows*).

B: ERCP confirmed the presence of a cystic duct remnant. Also noted was a slightly dilated bile duct without any evidence of stones (* = air bubble). This patient's symptoms were felt to be related to inflammation within the cystic duct remnant with the recent passage of retained stones.

(Reprinted from D'Alonzo W, Velchik MG. Post cholecystectomy syndrome due to a cystic duct remnant diagnosed by hepatobiliary scintigraphy. *Clin Nucl Med*. 1984;9:719. With permission.)

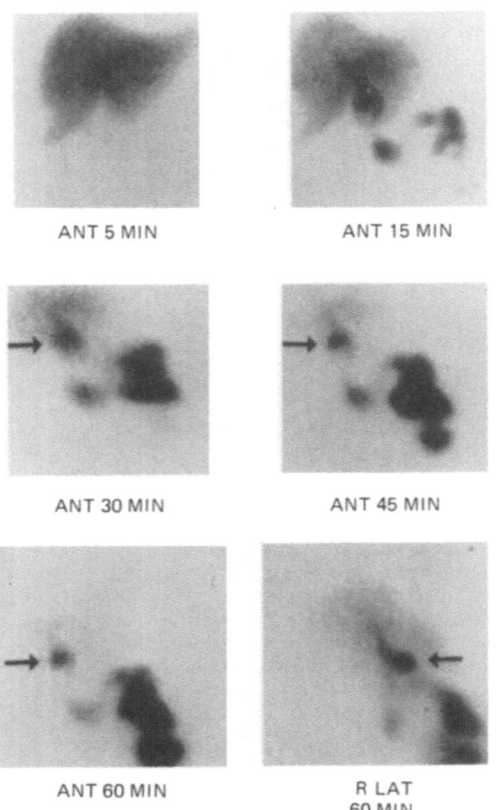

ANT 5 MIN   ANT 15 MIN

ANT 30 MIN   ANT 45 MIN

ANT 60 MIN   R LAT
                   60 MIN   **A**

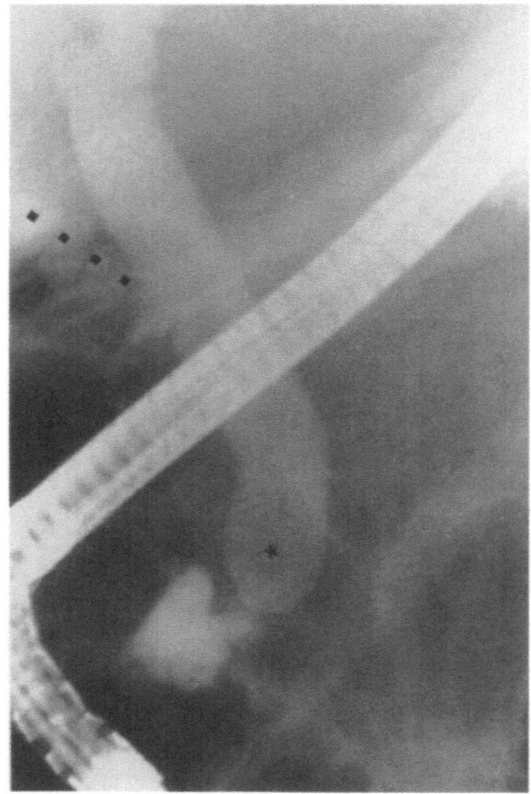

B

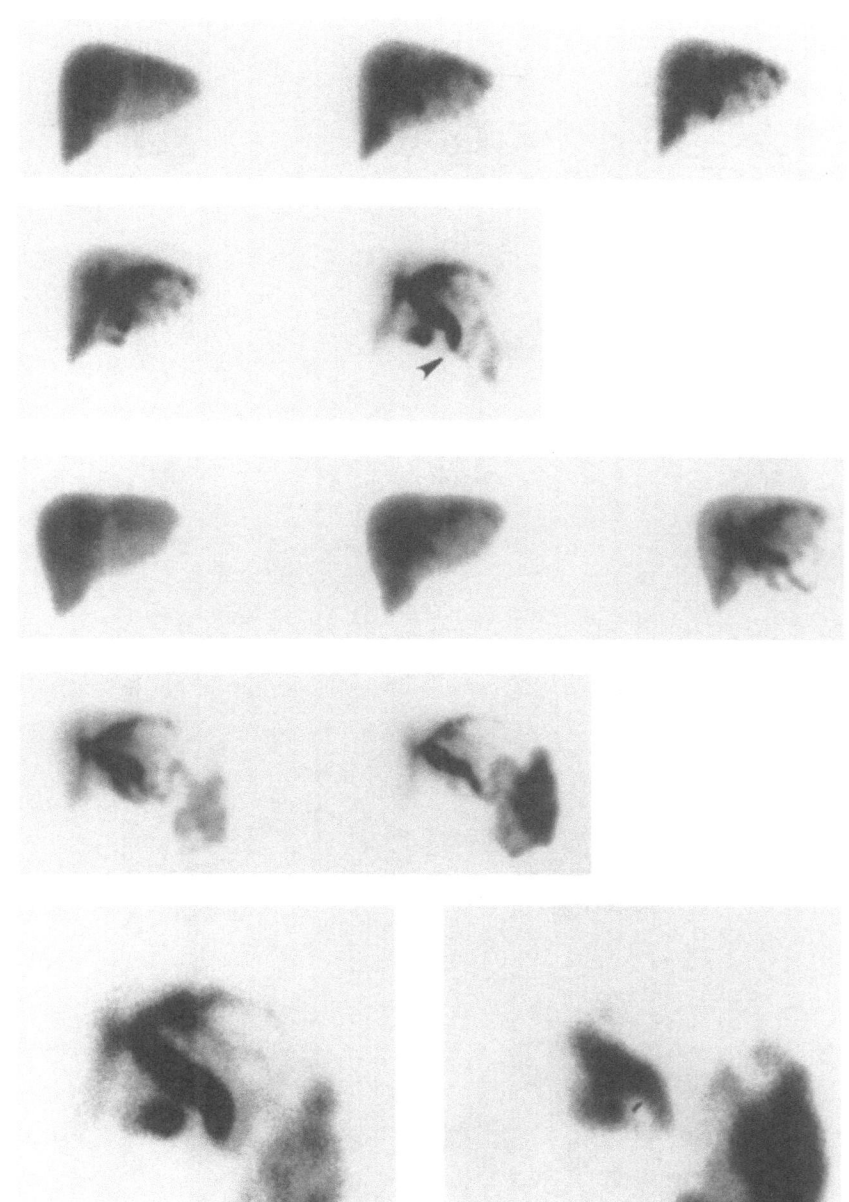

Figure 1.16. Postcholecystectomy biliary stricture.

This patient had recurrent abdominal pain 1 year after cholecystectomy.

A: Cholescintigraphy is consistent with partial common bile duct obstruction. Note the diffusely dilated intrahepatic ducts, dilated common hepatic, and common bile duct with apparent cutoff at the distal sphincter of Oddi. There is some biliary-to-bowel transit seen on the last image at 60 min (*arrowhead*). Localized duodenal activity is noted inferior to the common duct.

Comment: Biliary to bowel transit may be normal (seen within one hour) in partial common duct obstruction.

B: Repeat cholescintigraphy performed with constant cholecystokinin infusion, 0.04 μg/kg over 60 min. Although there still is an abrupt cutoff at the sphincter and dilated common duct, biliary-to-bowel transit is increased. ERCP diagnosed a postinflammatory stricture of the distal common bile duct. Sphincter of Oddi basal pressure was 17 mm Hg (normal). Sphincterotomy was performed.

C: Improved common duct clearance between 1 (*left*) and 2 (*right*) hr. This is not what one expects to see in a significant obstruction. Usually activity within the duct does not lessen but rather increases. This patient did symptomatically improve with sphincterotomy.

Comment: This case illustrates the difficulty of using any one criterion alone or requiring all the usual criteria in order to diagnose partial common duct obstruction. CCK infusion is not suggested here as a diagnostic test, but was part of an ongoing research study of sphincter dysfunction. However, it does demonstrate that even with a fixed obstruction there may be some physiological reversibility.

Figure 1.17. Sphincter of Oddi dysfunction: papillary stenosis.

This fifty-year-old woman was 3 yrs postcholecystectomy with chronic recurrent pain. ERCP showed no mechanical obstruction but basal sphincter of Oddi pressure was elevated (45 mm Hg). Preoperative (**A**) and postoperative (**B**) ^{99m}Tc disofenin studies are shown. Sphincterotomy was successful.

A: Preoperative HIDA study shows delayed hepatobiliary clearance with retention of activity in the common bile duct (CD) at 60 min (*arrow*). Increasing activity within the duodenum (D) is noted. A second preoperative study (not shown here) with constant CCK infusion (0.04 μg/kg infused over 60 min) was not significantly different (i.e., this was a fixed papillary stenosis).

B: Postsphincterotomy HIDA study in the same patient as **A** shows significant improvement with rapid hepatobiliary and common duct clearance compared to the preoperative study. CD, common duct; D, duodenum; S, stomach.

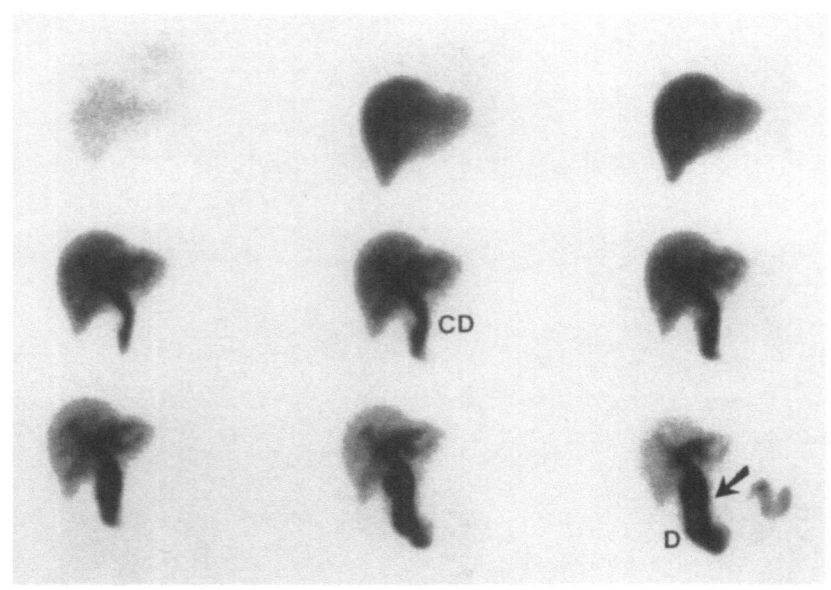

A

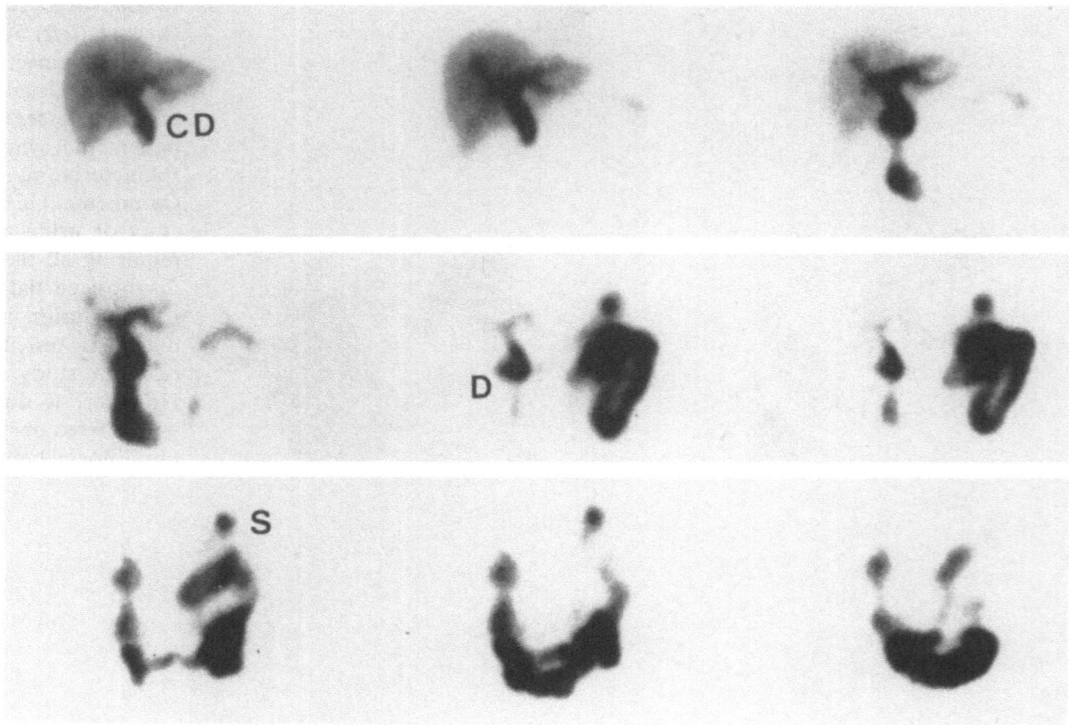

B

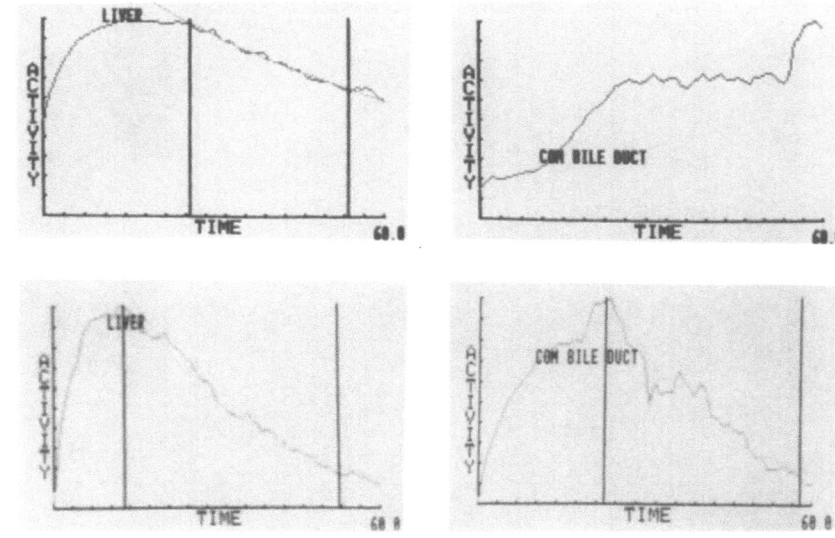

C: Preoperative (*above*) and postoperative (*below*) time–activity curves for two regions of interest: a) combined liver and common bile duct (*left*) and common bile duct alone (*right*). The time–activity curves and resulting quantitation showed significant postoperative improvement as demonstrated by an earlier time-to-peak and more rapid clearance slopes ($t\frac{1}{2}$). Obviously, the quantitative parameters improved dramatically.

Figure 1.18. Sphincter of Oddi dysfunction: dyskinesis.

Postcholecystectomy pain syndrome. ERCP showed no mechanical obstruction. Sphincter of Oddi manometry measured basal sphincter of Oddi pressure of 48 mm Hg. A sphincterotomy was performed.

A: Preoperative study. Sequential analog images over 60 min shows a prominent intrahepatic collecting system with dilitation in the region of the common hepatic duct at 60 min. A delayed image at 2 hr shows that the obstruction is really at the level of the sphincter of Oddi (*arrowhead*).

B: Preoperative study with a continuous infusion of sincalide, 0.04 μg/kg, over 1 hr. Hepatobiliary clearance is more rapid than the study without CCK (Fig. 1.18A). However, at the end of 60 min there is still retained activity in a prominent common duct. This is an obstructed dyskinetic sphincter of Oddi.

C: Postsphincterotomy study. There is still prominent retention in the common duct, but hepatobiliary clearance has significantly improved since the baseline study (**A**) and, interestingly, looks similar to the preoperative study with CCK study (**B**).

D: Time–activity curves for a region of interest including both the liver and common bile duct for the patient in **A,B,C**. *Left:* Preoperative study showing delayed peak and no downslope to the clearance phase. *Middle:* Preoperative study with constant CCK infusion showing considerable improvement with much earlier time-to-peak (19 min) and definite downsloping clearance phase ($t_{\frac{1}{2}}$ of 21 min). *Right:* Postoperative curve showing time–activity curve almost as good as the preoperative study with CCK infusion (*middle*). Time-to-peak is 20 min and $t_{\frac{1}{2}}$ was 40 min.

E: Time–activity curves for a region of interest for the common bile duct for the patient in **A,B,C**. *Left:* Preoperative study shows only upsloping without peaking. *Middle:* Preoperative study with constant CCK infusion shows definite peak (31 min) and downsloping clearance curve ($t_{\frac{1}{2}}$ of 57 min). *Right:* Postoperative curve similar to preoperative study with CCK infusion (*middle*). Time-to-peak is 29 min and $t_{\frac{1}{2}}$ 19 min.

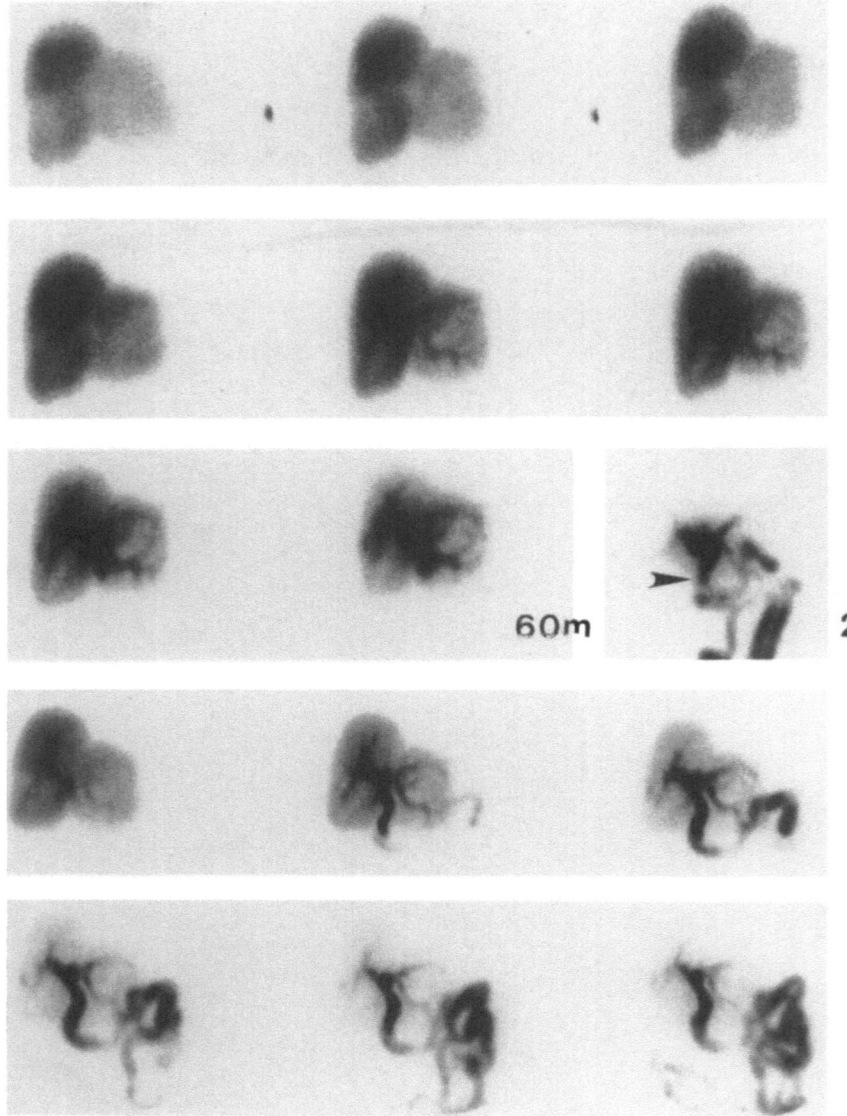

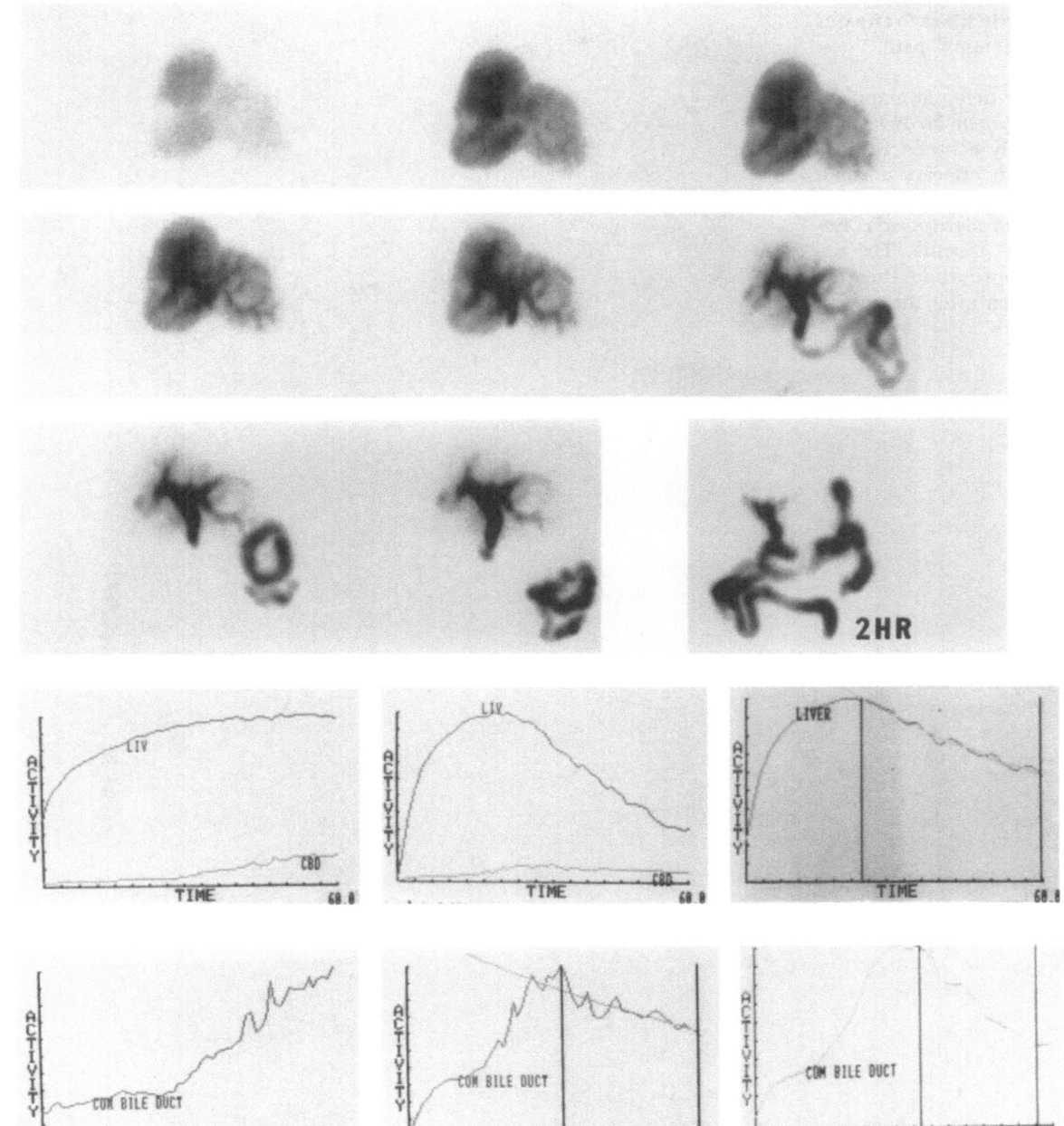

Figure 1.19. Enterogastric biliary reflux as cause of "postcholecystectomy" pain.

Although there is some delay in emptying of the common duct (*open arrowheads*), it does empty completely between 1 and 2 hr. Hepatobiliary transit appears normal. However, there is significant biliary reflux into the stomach (*closed arrowhead*). Endoscopy confirmed bile gastritis. The patient responded to appropriate therapy. The last image was taken after the patient ingested water.

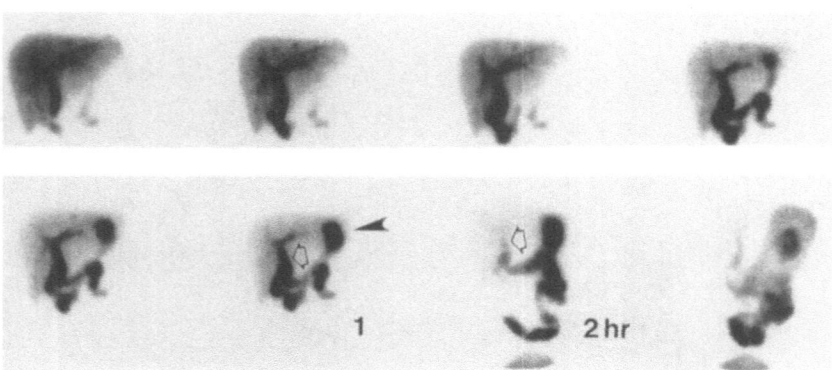

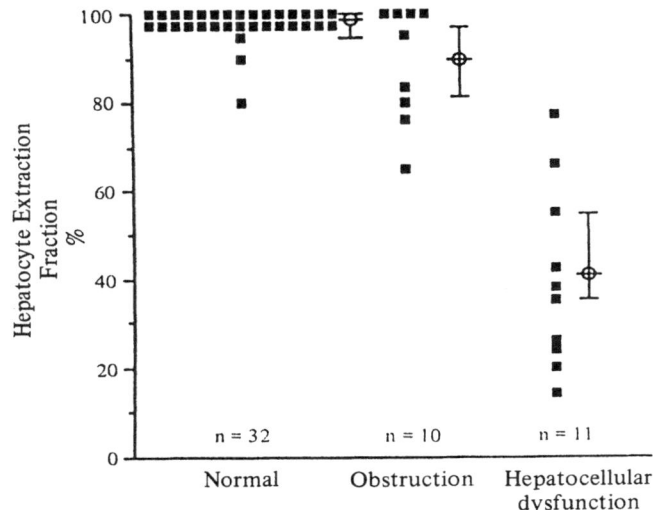

A

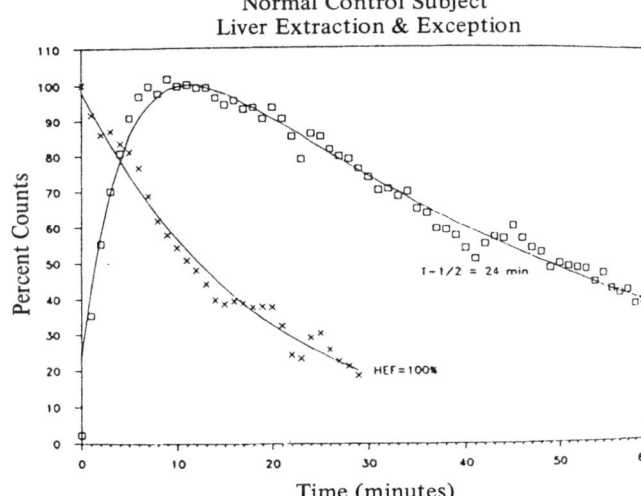

B

Figure 1.20. Deconvolutional analysis for differentiating hepatocyte dysfunction from primary biliary obstruction.

A: Hepatocyte extraction fraction (HEF) for normal controls (NC), isolated common bile duct obstruction (ICBDO), sclerosing cholangitis (SC), and alcoholic cirrhosis (CIRR). For ICBDO, P denotes partial obstruction and T denotes total obstruction. Child classification for CIRR: A (mild), B (moderate), or C (severe).

B: Normal control subject liver extraction and excretion. The X data points from 1 to 30 min are the deconvolved extraction data with the exponential fitted curve defining the HEF as noted on the figure. The square data points from 1 to 60 min represent the background corrected liver counts, normalized to 100% at the maximum. The fitted line 1 to 60 min is the two-compartment excretion model with an excretion $t_{\frac{1}{2}}$ as noted on the figure.

C: Same data as in **B** for isolated common bile duct obstruction. Sclerosing cholangitis gives similar results.

D: Same data as in **B** and **C** for alcoholic (Laennec's) cirrhosis.

(Reprinted from Brown PH, Juni JE, Lieberman DA, Krishnamurthy GT. Hepatocyte versus biliary disease: a distinction by deconvolutional analysis of technetium-99m IDA time-activity curves. *J Nucl Med.* 1988;29:623–630. With permission.)

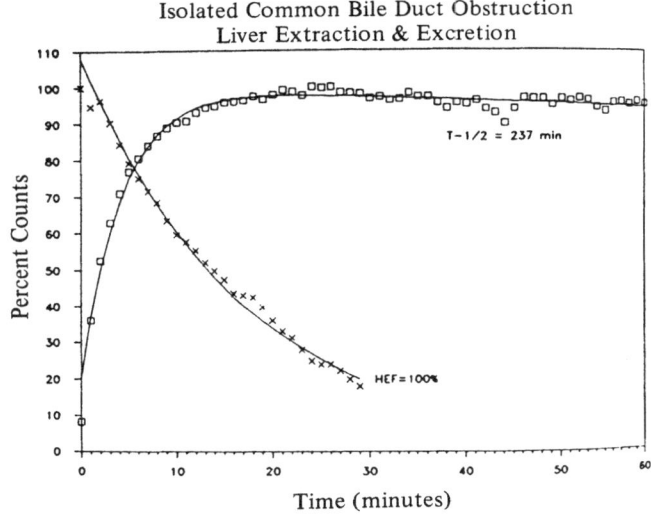

C

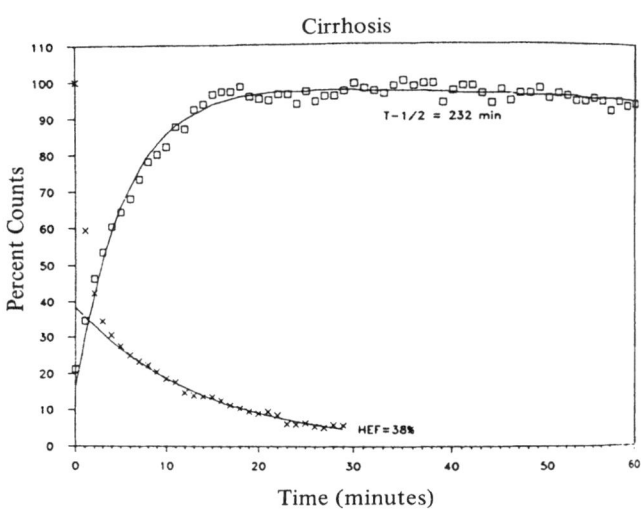

D

References

1. Blue PW. Hepatobiliary imaging in patients without prior surgery. In: Van Nostrand D and Baum S, eds. *Atlas of Nuclear Medicine*. Philadelphia: JB Lippincott Co; 1988:41–51.

2. Kuni CC, Klingensmith WC, eds. *Atlas of Radionuclide Hepatobiliary Imaging*. Boston: GK Hall Medical Publishers; 1983.

3. Weissmann HS, Sugarman LA, Freeman LM. Atlas of Tc-99m iminodiacetic acid (IDA) cholescintigraphy. *Clin Nucl Med*. 1982;7:231–239.

4. Weissman HS, Berkowitz D, Fox MS, Gliedman ML, Rosenblatt R, Sugarman LA, Freeman LM. The role of technetium-99m iminodiacetic acid (IDA) cholescintigraphy in acute acalculous cholecystitis. *Radiology*. 1983;146:177–180.

5. Krishnamurthy GT, Turner FE. Pharmacokinetics and clinical applications of technetium 99m-labeled hepatobiliary agents. *Sem Nucl Med*. 1990;20:130–149.

6. Chaudhuri TK, Fink S. Physiologic considerations in radionuclide imaging of the extrahepatic biliary tract. *Am J Phys Imaging*. 1988;3:114–120.

7. Weissmann HS, Badia J, Sugarman LA, et al. Spectrum of cholescintigraphic patterns in acute cholecystitis. *Radiology*. 1981;138:167–175.

8. Freitas JE, Coleman RE, Nagle CE, Bree RL, Kremer KD, Gross M. Influence of scan and pathological criteria on the specificity of cholescintigraphy. *J Nucl Med*. 1983;24:876–879.

9. Weissmann HS, Freeman LM. The biliary tract. In: Freeman LM, ed. *Freeman and Johnson's Clinical Radionuclide Imaging*, 3rd ed. New York: Grune & Stratton; 1984: 916–1049.

10. Kim CK, Palestro CJ, Solomon RW, Molinari DS, Lee SO, Goldsmith SJ. Delayed biliary-bowel transit in cholescintigraphy after cholecystokinin treatment. *Radiology*. 1990;176:553–556.

11. Drane WE. Gallbladder visualization after hepatic washout. *Clin Nucl Med*. 1987;12:453–456.

12. Choy D, Shi EC, McLean RG, et al. Cholescintigraphy in acute cholecystitis: use of intravenous morphine. *Radiology*. 1984;151:203–207.

13. Fig LM, Wahl RL, Stewart RE, Shapiro B. Morphine-augmented hepatobiliary scintigraphy in the severely ill: caution is in order. *Radiology*. 1990;175:467–473.

14. Kistler AM, Ziessman HA, Gooch D, Bitterman P. Utility of morphine-augmented hepatobiliary scanning in evaluation of acute cholecystitis. *Clin Nucl Med*. 1991;16:404–406.

15. Vasquez TE, Rimkus DS, Pretorius HT, Greenspan G. Intravenous administration of morphine sulfate in hepatobiliary imaging for acute cholecystitis: a review of clinical efficacy. *Nucl Med Commun*. 1988;9:217–222.

16. Kim EE, Pjura G, Lowre P, Nguyen M, Pollack M. Morphine-augmented cholescintigraphy in the diagnosis of acute cholecystitis. *AJR*. 1986;147:1177–1179.

17. Mock JM, Slavin JD Jr, Spencer RP. Two false negative results using morphine sulfate in hepatobiliary imaging. *Clin Nucl Med*. 1989;141:87–88.

18. Brachman MB, Tanasescu DE, Ramanna L, Waxman AD. Acute gangrenous cholecystitis: radionuclide diagnosis. *Radiology*. 1984;151:209–211.

19. Smith R, Rosen JM, Alderson PO. Gallbladder perforation: diagnostic utility of cholescintigraphy in suggested subacute or chronic cases. *Radiology*. 1986;158:63–66.

20. Meekin GK, Ziessman HA, Klappenbach RS. Prognostic value and pathophysiologic significance of the rim sign in cholescintigraphy. *J Nucl Med*. 1987;28:1679–1682.

21. Swayne LC, Ginsberg HN. Diagnosis of acute cholecystitis by cholescintigra-

phy: significance of pericholecystic hepatic uptake. *AJR*. 1989;152:1211–1213.

22. Bushnell DL, Perlman SB, Wilson MA, et al. The rim sign: association with acute cholecystitis. *J Nucl Med*. 1986;27:353–356.

23. Colletti PM, Cirimelli KM, Radin DR, et al. Radionuclide angiography in suspected acute cholecystitis: further observations. *Clin Nucl Med*. 1989;14:867–873.

24. Colletti PM, Ralls PW, Siegel ME, Halls JM. Acute cholecystitis: diagnosis with radionuclide angiography. *Radiology*. 1987;163:615–618.

25. Rogers JV, Mack L, Freeny PC, Johnson ML, Sones PJ. Hepatic focal nodular hyperplasia: angiography, TCT, sonography, and scintigraphy. *AJR*. 1981;137:983–990.

26. Tanasescu D, Brachman M, Rigby J, et al. Scintigraphic triad in focal nodular hyperplasia. *Am J Gastroenterol*. 1984;79:61–64.

27. Biersack HJ, Thelen M, Torres JF, et al. Focal nodular hyperplasia of the liver as established by 99mTc sulfur colloid and HIDA scintigraphy. *Radiology*. 1980;137:187–190.

28. Hasegawa Y, Nakano S, Ibuka T, et al. The importance of delayed imaging in the study of hepatoma with a new hepatobiliary agent. *J Nucl Med*. 1984;25:1122–1126.

29. Calvet X, Pons F, Bruix J, et al. Technetium-99m DISIDA hepatobiliary agent in diagnosis of hepatocellular carcinoma: relationship between detectability and tumor differentiation. *J Nucl Med*. 1988;29:1916–1920.

30. Kipper MS, Reed KR, Contardo M. Visualization of hepatic adenoma with Tc-99m di-isopropyl IDA. *J Nucl Med*. 1984;25:986–988.

31. Gratz KF, Creutzig H, Brolsch C, et al. Choleszintigraphie zum nachweis der focal-nodularen hyperplasie (FNH) der leber? *Chirurg*. 1984;55:448–451.

32. Kotzerke J, Schwarzrock R, Krischek O, Wiese H, Hundeshagen H. Technetium-99m DISIDA hepatobiliary agent in diagnosis of hepatocellular carcinoma, adenoma, and focal nodular hyperplasia. *J Nucl Med*. 1989;30:1278–1279.

33. Rodman CA, Keefe EB, Lieberman DA, et al. Diagnosis of sclerosing cholangitis with technetium 99m-labeled iminodiacetic acid planar and single photon emission computed tomographic scintigraphy. *Gastroentrology*. 1987;92:777–785.

34. Béla, Halpert. In: Anderson WHD, ed. *Pathology*, Chapter 29. St. Louis: CV Mosby Co; 1971:1261–1265.

35. Fink-Bennett D. The role of cholecystogogues in the evaluation of biliary tract disorders. In: Freeman LM, Weissmann HS, eds. *Nuclear Medicine Annual*. New York: Raven Press, 1985.

36. Swayne L, Palace F, Rothenberg J, et al. Sensitivity, specificity, and accuracy of CCK cholescintigraphy in acalculous biliary disease. *J Nucl Med*. 1986;27:882–883. Abstract.

37. Brugge WR, Brand DDL, Atkins JL, et al. Gallbladder dyskinesia in chronic acalculous cholecystitis. *Dig Dis Sci*. 1986;31:461–467.

38. Topper TE, Ryerson TW, Nora PF. Quantitative gallbladder imaging following cholecystokinin. *J Nucl Med*. 1980;21:694–696.

39. Westlake PJ, Hershfield NB, Kelly JK, et al. Chronic right upper quadrant pain without gallstones: does HIDA scan predict outcome after cholecystectomy? *Am J Gastroenterol*. 1990;85:986–990.

40. Davis GB, Berk RN, Scheible FW, et al. Cholecystokinin cholescintigraphy, sonography and scintigraphy: detection of chronic acalculous cholecystitis. *AJR*. 1982;1139:1117–1121.

41. Weissmann HS, Gliedman MR, Wilk PJ, et al. Evaluation of the postoperative patient with 99mTc-IDA cholescintigraphy. *Semin Nucl Med*. 1982;12:27–52.

42. Cruetzig H, Gratz K, Brolsch C, et al. Diagnosis of bile leakage by cholescintigraphy. *J Nucl Med*. 1982;23:73.
43. Rosenthall L, Fonseca C, Arzoumanian A, et al. 99mTc-IDA hepatobiliary imaging following upper abdominal surgery. *Radiology*. 1979;130:735-739.
44. D'Alonzo W, Velchik MG. Post cholecystectomy syndrome due to a cystic duct remnant diagnosed by hepatobiliary scintigraphy. *Clin Nucl Med*. 1984; 9:719.
45. Zeman RK, Lee C, Jaffe MH, Burrell JI. Hepatobiliary scintigraphy and sonography in early biliary obstruction. *Radiology*. 1984;153:793-798.
46. Zeman RK, Burrell MI, Dobbins J, Jaffe MH, Choyke PL. Postcholecystectomy syndrome: evaluation using biliary scintigraphy and endoscopic retrograde cholangiopancreatography. *Radiology*. 1985;156:787-792.
47. Krishnamurthy GT, Lieberman DA, Brar HS. Detection, localization and quantitation of degree of common bile duct obstruction by scintigraphy. *J Nucl Med*. 1985;26:726-735.
48. Kloiber R, AuCoin R, Hershfield NB, et al. Biliary obstruction after cholecystectomy: diagnosis with quantitative cholescintigraphy. *Radiology*. 1988; 169:643-647.
49. Darweesh RMA, Dodds WJ, Hogan WJ, et al. Efficacy of quantitative hepatobiliary scintigraphy and fatty-meal sonography for evaluating patients with suspected partial common duct. *Gastroenterology*. 1988;94:779-786.
50. Steinberg WM. Sphincter of Oddi dysfunction: a clinical controversy. *Gastroenterology*. 1988;95:1409-1415.
51. Toouli J, Roberts-Thomson, IC, Dent J, Lee J. Manometric disorders in patients with suspected sphincter of Oddi dysfunction. *Gastroenterology*. 1985:88:1243-1250.
52. Venu RP, Greenen JE. Diagnosis and treatment of diseases of the papilla. *Clin Gastroenterol*. 1986;15:439-455.
53. Shaffer EA, Hershfield NB, Logan HK, Kloiber R. Cholescintigraphic detection of functional obstruction of the sphincter of Oddi. Effect of papillotomy. *Gastroenterology*. 1986;90:728-733.
54. Lee RGL, Gregg JA, Koroshetz AM, Hill TC, Clouse ME. Sphincter of Oddi stenosis: diagnosis using hepatobiliary scintigraphy and endoscopic manometry. *Radiology*. 1985;156:793-796.
55. Pace RF, Chamberlain MJ, Passi RB. Diagnosing papillary stenosis by technetium-99m HIDA scanning. *Can J Surg*. 1983;26:191-193.
56. Hershfield NB. Diagnosis of sphincter of Oddi dysfunction. *Gastroenterology*. 1989;95:241.
57. Roberts-Thomson IC. Endoscopic sphincterotomy of the papilla of Vater: an analysis of 300 case. *Aust NZ J Med*. 1984;14:611-617.
58. Geenen JE, Hogan WJ, Dodds WJ, et al. The efficacy of endoscopic sphincterotomy after cholecystectomy in patients with sphincter-of-oddi dysfunction. *N Engl J Med*. 1989;320:82-87.
59. Carr-Locke DL, Bailey I, Neoptolemos J, Leese T, Heath D. Outcome of endoscopic sphincterotomy for papillary stenosois. *Gut*. 1986;27:A1280.
60. Brough WA, Taylor TV, Torrance HB. The effect of cholecystectomy on duodenogastric reflux in patients with previous peptic ulcer surgery. *Scand J Gastroenterol*. 1984;19;255-256.
61. Coletti RM, Barakos JA, Siegel ME, et al. Enterogastric reflux in suspected acute cholecystitis. *Clin Nucl Med*. 1987;12:533-535.
62. Drane WE, Karvelis K, Johnson DA, Silverman ED. Scintigraphic evaluation of duodenogastric reflux. Problems, pitfalls and technical review. *Clin Nucl Med*. 1987;12:377-384.
63. Brown PH, Juni JE, Lieberman DA, Krishnamurthy GT. Hepatocyte versus biliary disease: a distinction by deconvolutional analysis of technetium-99m IDA time-activity curves. *J Nucl Med*. 1988;29:623-630.

CHAPTER 2

Atlas of Gastrointestinal Bleeding (RBC) Scintigraphy

Philip W. Wiest and Michael F. Hartshorne

The detection and location of gastrointestinal (GI) bleeding has become an important part of routine nuclear medicine practice. This study must be understood, done well, and read skillfully in order to provide a service to the referring physician. Much of the confusion over the reading of these scans can be readily explained as a failure to follow the rules of interpretation. Many of the published "pitfalls" are really examples of abnormal findings or normal variants that can be readily separated from the studies that successfully localize hemorrhage.[1] It is also important to point out that scintigraphy is not done to determine that there *is* hemorrhage. That much is usually known at the start. No one needs a "nuclear guaiac." Localization of the site of hemorrhage is paramount in guiding further evaluation and therapy. This discussion will explain the technique and interpretation of red blood cell (RBC) GI bleeding study.

Red Blood Cell Labeling Technique

The goal of any GI bleeding study, to locate the site of hemorrhage, demands a method of RBC labeling that will prevent false positive interpretations caused by physiologic secretion of technetium-99m pertechnetate ($^{99m}TcO_4$) by the stomach. The early in vivo method of Pavel was technically easy. It consisted of an injection of stannous pyrophosphate followed by a second injection of $^{99m}TcO_4$.[2] Unfortunately, this method was handicapped by the excretion of free $^{99m}TcO_4$ by the stomach and kidneys. Nasogastric suction as well as intravenous sodium perchlorate or cimetidine have been employed to increase test specificity by reducing false positive bleeding studies.[3,4] However, these extra steps are inadequate to prevent all gastric secretion.

To reduce the extravascular $^{99m}TcO_4$, in vitro methods were developed. The Brookhaven Laboratory method is the best known.[5,6] This kit uses a reaction vial in which packed RBCs are treated with stannous ion and then labeled with $^{99m}TcO_4$. Labeled RBCs are then separated and washed, if necessary, before injection. This method produces high labeling efficiency (95% +) but is time consuming and not commercially available.

Modified techniques were developed by multiple researchers to bridge the gap between easy but inadequate in vivo methods and efficient but investigational in vitro methods. The modified in vitro (commonly known as the "invivtro" technique) deserves description. Stannous ion is administered intravenously as stannous pyrophosphate (known by the

Table 2.1. Method of RBC labeling according to Benedetto and Nusynowitz.[9]

1. Prepare and inject stannous pyrophosphate according to instructions on pyrophosphate kit
2. After 15 min collect 6 ml of blood into heparinized blood tube
   SHIELD REMAINING STEPS:
3. Inject 20–30 mCi (74–111 MBq) (adult dose) $^{99m}TcO_4$ into the collection tube containing the blood
4. Mix gently and incubate at room temperature for 5 min
5. Centrifuge collection tube stopper down for 5 min (Note: protective sleeve is necessary to maintain vertical alignment of tube in most centrifuges)
6. Gently remove the tube and keep inverted
7. Using a 10-ml syringe with attached 21-ga 1.5″ needle, insert needle to the hub. Avoid disturbing RBC layer until the needle is above cell layer
8. Aspirate the entire plasma layer
9. Withdraw needle from blood tube while maintaining a firm grip on the plunger to prevent syringe contents from being inadvertently sucked back into the vacuum tube
10. Approximately 2 ml of RBCs will remain. Add 2–4 ml of sterile saline and invert tube several times to suspend cells
11. Place tube carefully into a dose calibrator to determine total activity
12. Withdraw necessary activity for the patient dose and inject

slang term "cold pyro"). After this, blood is drawn into a heparinized syringe containing the dose of $^{99m}TcO_4$ and allowed to incubate for a 10-min period with gentle inversion of the syringe every minute for mixing. The contents of the syringe are then injected. The success of this procedure depends on the patient's hematocrit and the duration of the incubation. With a hematocrit of less than 30%, substantially longer incubation times are required. Callahan reported that 90% of cells were labeled using the recommended 10-min incubation time.[7] This was not duplicated by Landry, who determined that 45 min were required to achieve a 90% labeling of the RBCs[8] (Fig. 2.1). This discrepancy can probably be explained by differences in reaction kinetics due to dose. Callahan used microcuries of $^{99m}TcO_4$ in his study while Landry used a diagnostic dose of millicuries of $^{99m}TcO_4$.

Benedetto and Nusynowitz developed the optimal method for labeling RBCs.[9] Their technique is simple, has excellent labeling efficiency, and requires no special licensure (Table 2.1). In addition, their method is superior to the other methods because any free, untagged $^{99m}TcO_4$ is disposed of before injection. This alleviates reported problems such as poor tagging caused by numerous medications, blood transfusions, iodinated contrast, circulating antibodies, or a low hematocrit.[5] As this method may be unfamiliar, the authors recommend that it be practiced in conjunction with routine cardiac gated blood pool studies. Benedetto and Nusynowitz's labeling technique can be learned rapidly and employed with confidence when a GI bleeding study is required.

Acquisition

After preparation of the $^{99m}TcO_4$ RBCs the patient is positioned supine under the gamma camera *before* injection. The patient is then injected while slow dynamic (1 min) computer images with a 128 × 128 × 8 matrix are acquired. Byte mode frames should have sufficient pixel

Table 2.2. Computer acquisition of GI bleeding study.

Computer requirements
1. Matrix: 128 × 128 × 8
2. Zoom: none
3. Time/frame: 1 min
4. Number of frames: 90

Gamma camera requirements
1. Examination performed in nuclear medicine department: large field of view with a low energy parallel hole collimator
2. Examination performed mobile at the bedside: small field of view with a diverging low energy collimator

Film
1. One image every 5 min for 90 min

Clarification techniques
1. Cine (fast) computer replay
2. Lateral images to triangulate stationary "hot spots"
3. Void bladder to clear excreted isotope (Foley catheter may be necessary)
4. Void bowel to confirm low sigmoid/rectal/hemorrhoid bleeds
5. Correlate stationary "hot spots" with other imaging studies
6. If no bleed localized, consider 24-hr quality assurance image to check for "missed" bleeds

Repeat study protocol
1. Intermittent scanning possible for up to 24 hr after initial RBC label
2. Voiding bladder and bowel before repeat study may be helpful
3. Repeat computer/film acquisition for a minimum of 30 min
4. If there is extravasated $^{99m}TcO_4$ RBCs throughout bowel, localization may not be possible

depth. This computer acquisition is continued for a 90-min period and supplemented by an analog film image every 5 min (Table 2.2).

It is important to use a computer. Continuous sequential images permit cine playback, which allows the detection of small movements against the background noise and aids in the interpretation of subtle hemorrhage. The 90-min study period represents a practical cutoff point. Bunker demonstrated detection of 83% of active GI bleeds in this time frame[10] (Fig. 2.2).

Delayed images may be beneficial if the initial study was negative and there is evidence that the patient has rebled in the interim. If delayed imaging is required it may be performed up to 24 hr after the initial RBC label. Again, it is mandatory to acquire sequential computer images. These help differentiate between radiotracer activity extravasating at a more proximal or distal site and activity that has moved from the area of bleeding via peristalsis. If a significant time interval has occurred between initial and repeat imaging it may be discovered that the patient has hemorrhaged in the interim and the bowel has filled with $^{99m}TcO_4$ RBCs. This greatly hinders detecting the site of hemorrhage but is a useful quality assurance tool. The authors recommend obtaining a routine 24-hr image on all "negative" studies to find "missed" (nonlocalized) bleeds. If all the positive studies from the 90 min studies are added to all the delayed studies that find (but do not localize) blood in the bowel, a denominator representing total bleeds is obtained. This can be compared with a numerator listing all bleeds successfully localized. Thus:

Percent localized bleeds = 100[(localized bleeds) / (all bleeds)]

Using this method, a nuclear medicine service can create and maintain quality assurance records of its effectiveness in bleed detection and localization.

Interpretation

Diagnosis of GI bleeding depends on the interpreter's ability to detect extravasated, intraluminal $^{99m}TcO_4$ RBCs. Table 2.3 lists necessary scintigraphic criteria to identify bleeding accurately. An area of abnormal radiotracer activity (the "hot spot") appears from nowhere and conforms to the expected bowel anatomy. This activity should persist and may increase in overall intensity depending on the rate of hemorrhage. It is critical to the diagnosis that this activity move with peristalsis either antegrade or retrograde from the origin. Illustrations of positive GI scintigraphy are shown in Figs. 2.5 to 2.26. False positive interpretations can be avoided as long as careful attention is paid to these criteria. Pitfalls in the interpretation of GI scintigraphy are listed in Table 2.4 with illustrations in Figs. 2.28 to 2.50.

Thorne et al.[11] used dogs with a catheter inserted into the sigmoid lumen to demonstrate detection and localization of GI bleeding rates as low as 0.04 ml/min. This bleeding rate was visualized 55 to 63 min after the initiation of simulated hemorrhage. Experimental bleeding rates from 4.6 to 0.2 ml/min were all detected by 10 min. They concluded that a minimum of 2 to 3 ml of extravasated $^{99m}TcO_4$ RBCs was necessary for scintigraphic detection.

Smith et al.[12] calculated the mean bleeding rate in a series of 62 patients. They divided the patient's transfusion volume requirements by the reported duration of active bleeding and calculated that the minimum mean bleeding rate detectable by scintigraphy was 0.1 ml/min. This is a reasonable order of magnitude estimate. It is not directly comparable to Thorne's dog experiment since that study examined instantaneous bleeding rates instead of mean bleeding rates.

Multiple authors have stressed the importance of proper timing of the bleeding study. Patients should be referred to nuclear medicine early in the course of their GI hemorrhage and during a period of maximum blood loss. In a series of 100 patients, Winzelberg showed that 81% (39/48) of patients with a transfusion requirement of greater than one unit in the 24-hr period before scintigraphy had positive scans. Only 44% (23/52) of patients with transfusion requirements of one unit or less had

Table 2.3. Criteria for positive GI hemorrhage scintigraphy.

1. Abnormal radiotracer "hot spot" appears and conforms to bowel anatomy
2. Persistence of the abnormal activity (depending on the rate of hemorrhage, may become more intense with time)
3. Absolutely essential to the diagnosis: movement of the activity by peristalsis either retrograde or antegrade

Short version:
1. Comes outta nowhere
2. Gets hotter
3. Moves away through bowel

Table 2.4. GI bleed pitfalls in interpretation.

Common
 1. Physiologic: stomach, small intestine, bowel (Fig. 2.27)
 2. Physiologic: renal visualization
    a. Renal pelvic activity
    b. Pelvic kidney
    c. Ectopic kidney
Uncommon
 1. Hepatic hemangioma (Fig. 2.43)
 2. Varices (Figs. 2.35, 2.37, 2.39)
 3. Ureter (Fig. 2.28)
Rare
 1. Urinary bladder
 2. Abdominal aortic aneurysm (Fig. 2.32)
 3. Gastroduodenal artery aneurysm
 4. Pseudoaneurysms (Fig. 2.34)
 5. Uterine blush
 6. Male genitalia (Figs. 2.29, 2.30)
 7. Cutaneous hemangioma
 8. Leiomyosarcoma (Fig. 2.42)
 9. Angiomyolipomas of tuberous sclerosis
10. Abscess
11. Caput medusae and dilated mesenteric veins (Fig.2.36)
12. Ovarian vein (Fig. 2.38)
13. Gallbladder (Fig. 2.44)
14. Gluteal hematoma (Fig. 2.49)
15. Arterial grafts
16. Nonhemorrhagic gastritis (Fig. 2.31)
17. Factitious gastrointestinal bleeding (Fig. 2.25)

abnormal scans.[13] Smith et al. presented similar findings. Seventy-eight percent (18/23) of patients requiring two or more units (>500 ml) of blood in the 24-hr period preceding scintigraphy had positive scans. Only 33% (13/39) of patients with transfusion requirements of less than two units had abnormal scans. They concluded that the optimum time to refer patients to nuclear medicine for a GI bleeding study is during or immediately after a short period (<24 hr) in which the patient required at least a two-unit blood transfusion.[12]

Bunker et al. showed the anatomic distribution of 37 true-positive bleeding studies (Fig. 2.3). Most of the localized bleeds in his patient population occurred in the colon, with almost 25% of the bleeding diagnosed in the distal sigmoid colon or rectum.[10] Bunker and Hartshorne have suggested a clinical flow diagram for the evaluation of GI hemorrhage with $^{99m}TcO_4$ RBCs[14] (Fig. 2.4).

Dosimetry

The dosimetry of $^{99m}TcO_4$ RBCs is favorable and the whole body radiation dose is relatively low[5] (Table 2.5).

Table 2.5. Radiation absorbed dose from ^{99m}Tc-labeled RBCs.

Organ	Dose (rad/mCi)	mGy/MBq
Whole body	0.019	.005
Heart	0.078	.021
Spleen	0.050	.013
Liver	0.070	.019
Blood	0.055	.015
Lungs	0.056	.015
Kidneys	0.054	.015
Red marrow	0.033	.009

Modified from ref. 5.

Atlas Section

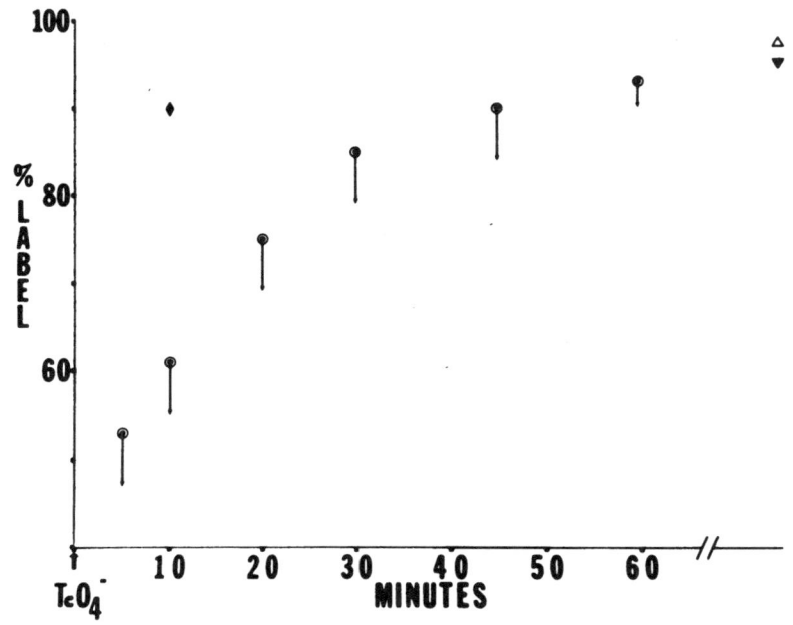

Figure 2.1. Kinetics of ^{99m}TcO$_4$-RBC labeling.

Shown is the increasing labeling efficiency of the "invivtro" technique (*circled dots, arrow* = 1 SD) with time. Note that the average labeling efficiency is finally 90% at 45 min. The 90% label reported by Callahan at 10 min is shown by a solid diamond. Pure in vivo methods are noted by the open and solid triangles.[8]
(Reprinted from Landry A, Hartshorne MF, Bunker SR, et al. Optimal technetium-99m RBC labeling for gastrointestinal hemorrhage study. *Clin Nucl Med*. 1985; 10:491–493. With permission.)

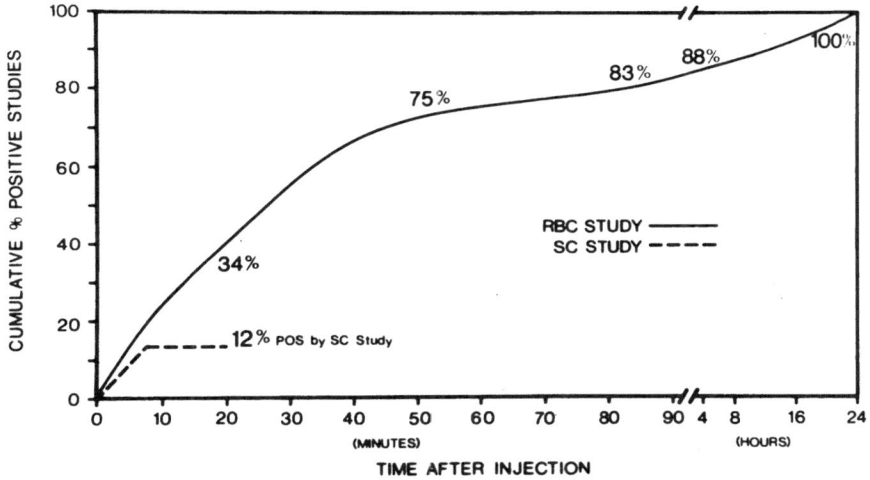

Figure 2.2. Times to positivity.

In 38 true positive scintigrams 83% of positive studies occurred by 90 min; mean time to positivity was 26 min. RBC = ^{99m}TcO$_4$ RBC; SC = ^{99m}Tc Sulfur colloid.[10]
(Reprinted from Bunker SR, Lull RJ, Tanasescu DE, et al. Scintigraphy of gastrointestinal hemorrhage: superiority of ^{99m}Tc red blood cells over 99mTc sulfur colloid. *AJR*. 1984;143:543–548. With permission.)

Figure 2.3. Location and frequency of GI bleeds.

The anatomic distribution and number of 37 true positive GI hemorrhage scintigrams shown is typical. One additional study had pancolonic activity secondary to excessive anticoagulant therapy.[10]
(Reprinted from Bunker SR, Lull RJ, Tanasescu DE, et al. Scintigraphy of gastrointestinal hemorrhage: superiority of ^{99m}Tc red blood cells over 99mTc sulfur colloid. *AJR*. 1984;143:543–548. With permission.)

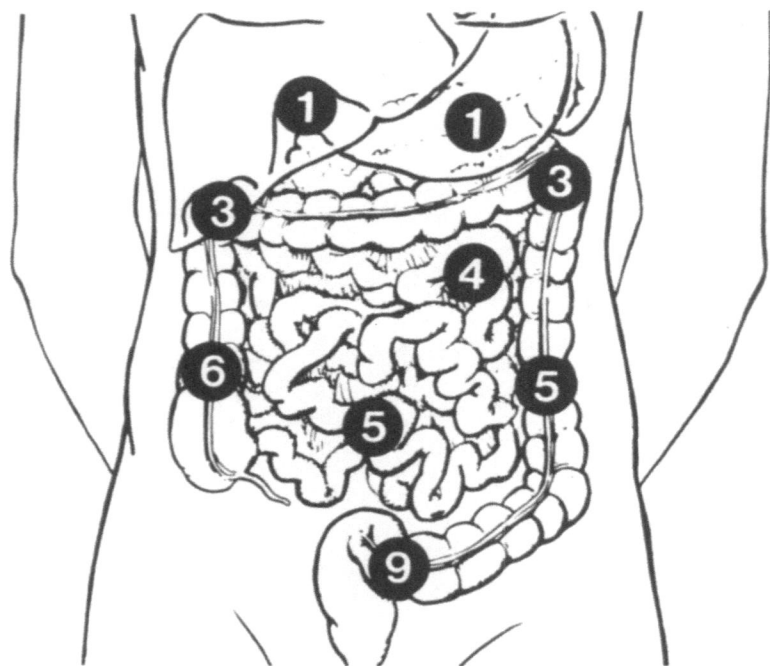

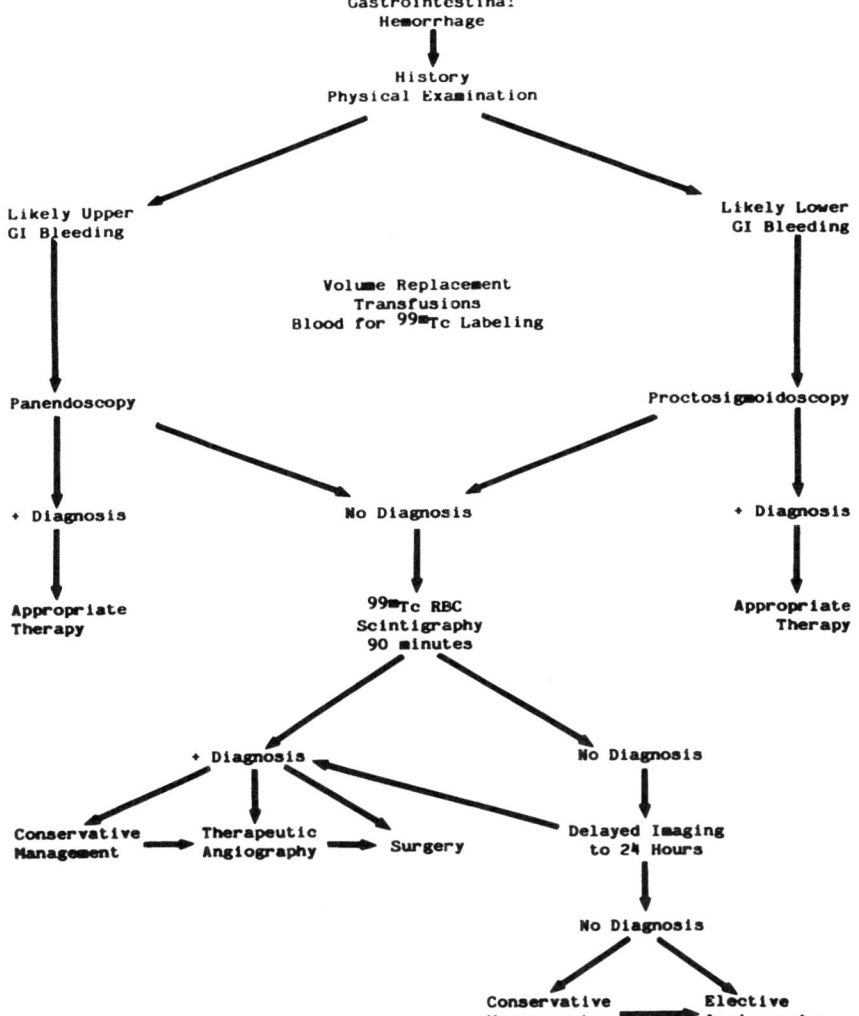

Figure 2.4. Flow chart for the evaluation of GI bleeding patients.[14]

(Reprinted from Bunker SR, Hartshorne MF. Gastrointestinal hemorrhage. In: Mettler, FA ed. *Radionuclide Imaging of the GI Tract*. New York: Churchill Livingstone; 1986: 53–81. With permission.)

Figure 2.5. Hemorrhage—gastric leiomyoma.

Free pertechnetate was excluded. **A:** $^{99m}TcO_4$ RBC image showing the immediate anterior abdomen. **B:** $^{99m}TcO_4$ RBC image showing the anterior abdomen 60 min after the injection. **C:** $^{99m}TcO_4$ RBC image showing the anterior abdomen 2 hr after injection. Note intense stomach activity (*arrows*).[15]
(Reprinted from Joseph UA, Jhingran SG. Technetium 99m-labeled RBC imaging in gastrointestinal bleeding from gastric leiomyoma. *Clin Nucl Med.* 1988;13:23–25. With permission.)

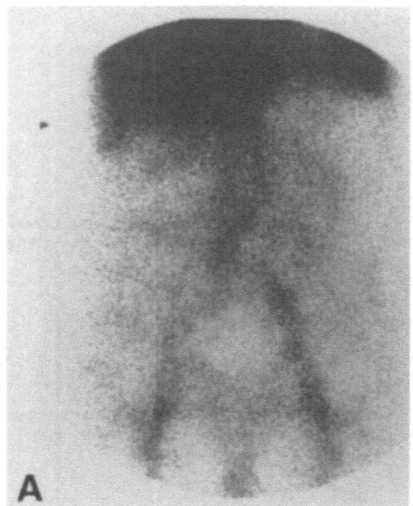
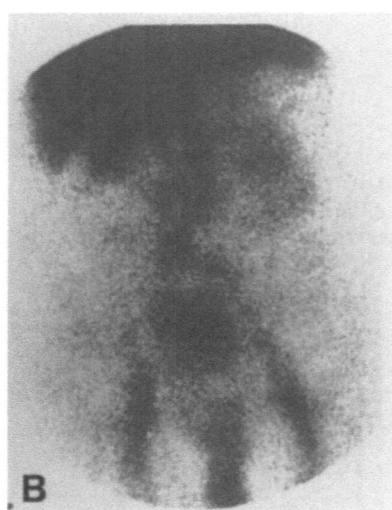
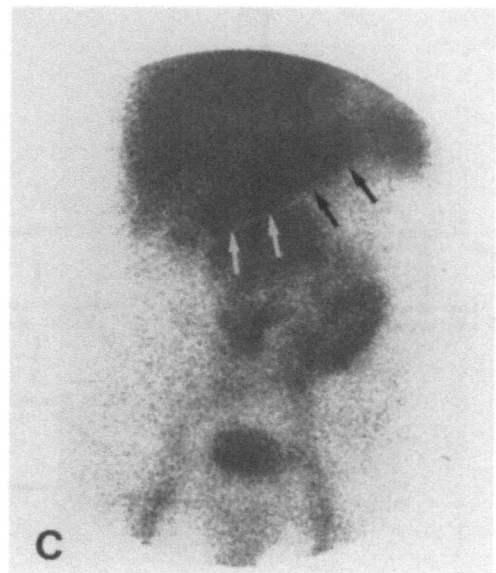

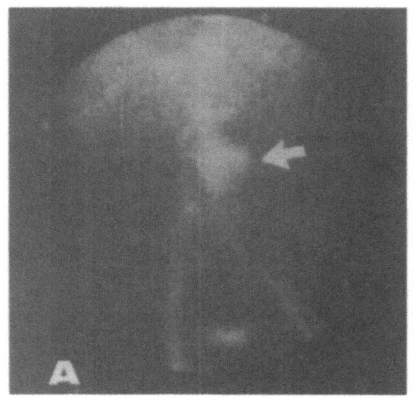
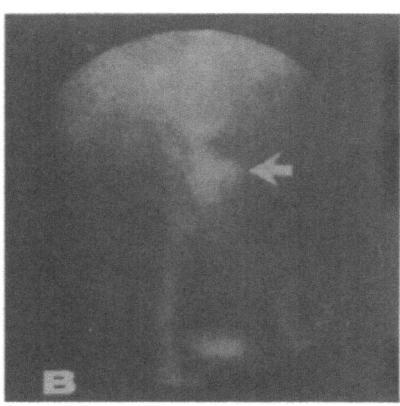
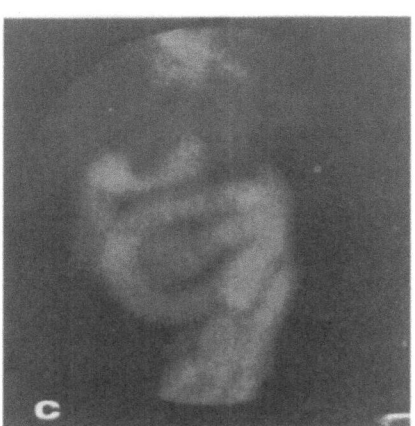

Figure 2.6. Hemorrhage — aortic aneurysm into duodenum.

Anterior abdominal images taken at 1 min (**A**) and 60 min (**B**) after $^{99m}TcO_4$ RBC injection. One-minute image shows radiotracer in abdominal vasculature, liver, and bladder. Patient's known abdominal aortic aneurysm is clearly visualized (*arrows*), represented by irregular and relatively intense radiotracer activity located just cephalad of the iliac bifurcation. Image taken at 60 min indicates essentially no change in radiotracer distribution. **C**: Image at 90 min after injection. Note the intense activity throughout the stomach, duodenum, and proximal small bowel. Patient's clinical course rapidly deteriorated, necessitating surgery. At operation it was found the patient's aneurysm had ruptured into the third portion of the duodenum.[16]
(Reprinted from Yen CK, Pollycove M, Parker H, Nalls G. Rupture of spontaneous aortoduodenal fistula visualized with Tc-RBC scintigraphy. *J Nucl Med.* 1983;24:332–333. With permission.)

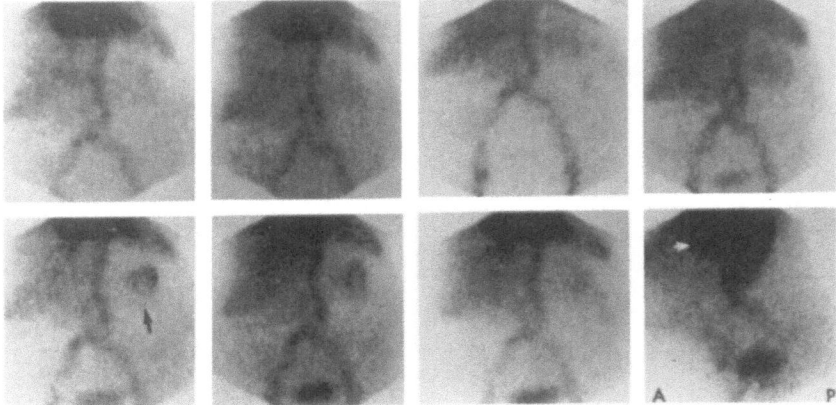

Figure 2.7. Angiodysplasia — jejunum.

Top row images taken (*from left to right*) at 5, 10, 15, and 30 min; bottom row images taken (*from left to right*) at 45, 60, 90, and 100 min. The 100-min image is a left lateral view (A, anterior; P, posterior). Note the extravasated radioactivity in the LUQ (*black arrow*), which is located anteriorly on the lateral view (*white arrow*). The exact site of the bleeding was difficult to establish. Cinematic playback demonstrated a pattern and direction of movement suggestive of small bowel loops. Surgery confirmed a jejunal bleeding source from angiodysplasia.[17]
(Reprinted from Tumeh SS, Parker JA, Royal HD, Uren RF, Kolodny GM. Detection of bleeding from angiodysplasia of the jejunum by blood pool scintigraphy. *Clin Nucl Med.* 1983;8:127–128. With permission.)

Figure 2.8. Hemorrhage—carcinoids small bowel.

An adult male patient had repeated GI hemorrhage and multiple negative upper and lower GI endoscopic and barium examinations. During a hypotensive episode, a $^{99m}TcO_4$ RBC study was performed. The images displayed here are 1 min frames, which begin at 60 min. The bladder activity (B) continues to accumulate as midline abnormal activity appears and then moves through twists and turns of mid to small bowel. At surgery the bleeding site was identified as two adjacent carcinoid tumors in the midjejunum.[14]
(Reprinted from Bunker SR, Hartshorne MF. Gastrointestinal hemorrhage. In: Mettler, FA ed. *Radionuclide Imaging of the GI Tract.* New York: Churchill Livingstone; 1986: 53–81. With permission.)

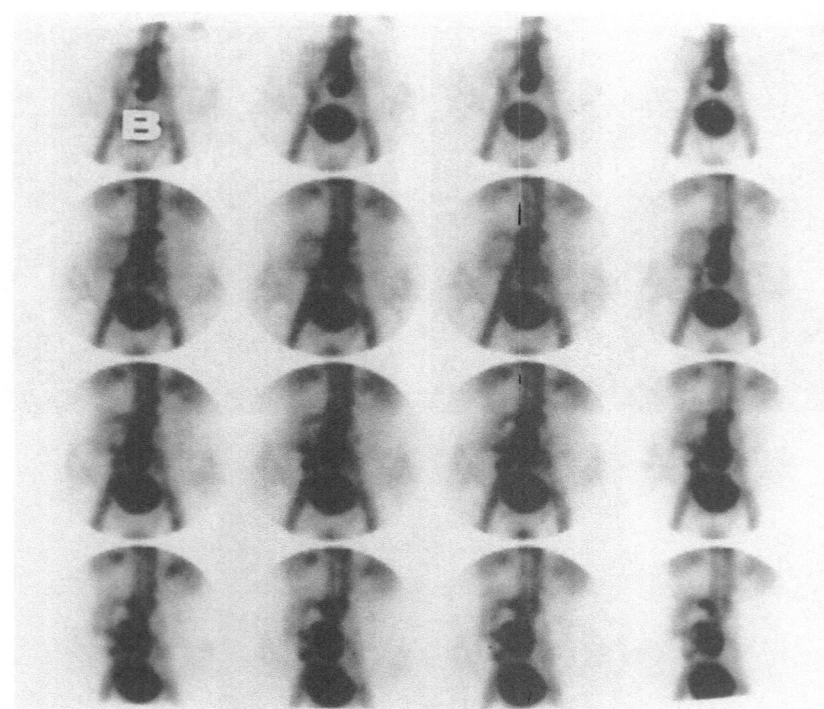

Figure 2.9. Hemorrhage—jejunum.

This example of distal jejunal bleeding has an unusual etiology. A $^{99m}TcO_4$ RBC scan on a 42-year-old man admitted for hematochezia 1 hr after accidentally being struck with a staple from an industrial staple gun.
A: The scan demonstrates extravasation of tracer into the GI tract with a pattern suggesting active small bowel bleeding. Forty minutes later much of the colon is outlined with tracer, suggesting continued extravasation.
B: Immediately thereafter, subsequent to a bowel movement, the tracer is seen in the rectosigmoid area.
C: Shortly thereafter a distal jejunal bleed was noted in the operating room and was repaired.[18]
(Reprinted from Schneider M, Siegel ME. Gastrointestinal bleeding from an unusual etiology. *Clin Nucl Med.* 1984;9:243. With permission.)

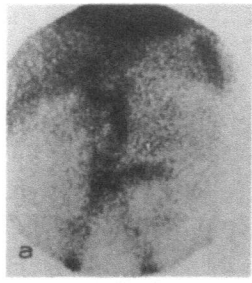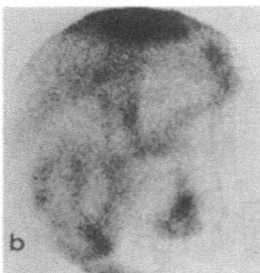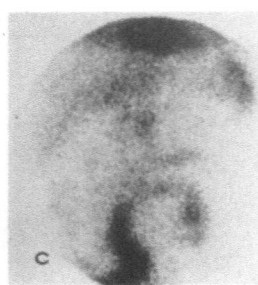

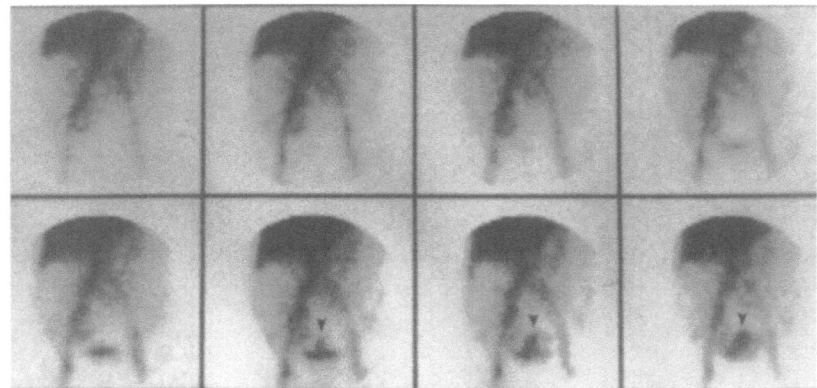

Figure 2.10. Hemorrhage — terminal ileum.

Selected 2 min $^{99m}TcO_4$ RBC computer images. Progressive hemorrhage into distal ileum (*arrowheads*). At surgery a large bleeding varix was found with the terminal ileum adherent to the bladder.[10] Having patient empty bowels and bladder as well as a lateral view would help confirm a location.
(Reprinted from Bunker SR, Lull RJ, Tanasescu DE, et al. Scintigraphy of gastrointestinal hemorrhage: superiority of ^{99m}Tc red blood cells over 99mTc sulfur colloid. *AJR*. 1984;143:543–548. With permission.)

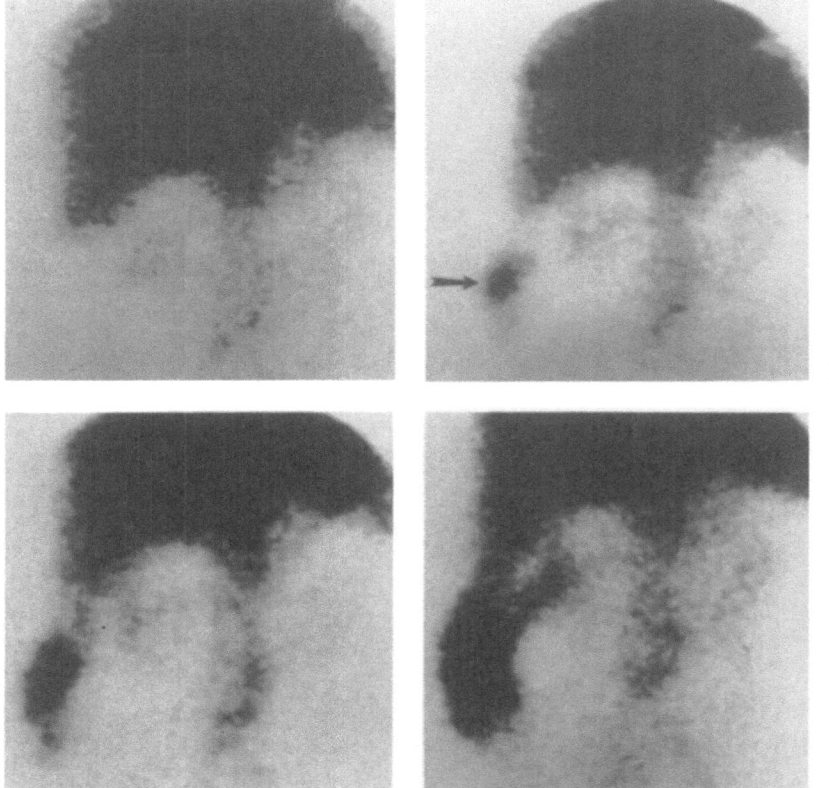

A   B   C   D

Figure 2.11. Hemorrhage — cecum.

Sequential anterior abdominal $^{99m}TcO_4$ RBC images. Ten (**A**), 20 (**B**), 25 (**C**), and 30 (**D**) min after injection. Abnormal focus of activity in cecum (*arrow*) is shown, which appears to be more intense than hepatic activity. Mean bleeding rate was 0.5 ml/min, and a right hemicolectomy was required for treatment.[12]
Comment: This case demonstrates well the criteria for a positive gastrointestinal scintigraphy (see Table 2.4).
(Reprinted from Smith R, Copley DJ, Bolen FH. 99mTc RBC scintigraphy: correlation of bleeding rates with scintigraphic findings. *AJR*. 1987;148:869–874. With permission.)

Figure 2.12. Indolent hemorrhage — cecum.

Anterior abdominal scintiscan of this elderly patient is normal at 1 hr. At 17 hr an abnormal focal collection is noted in the cecum (*arrow*). This activity may have flowed downstream from a more proximal site and pooled in the cecum so there is some uncertainty in localization. Surgery confirmed a bleeding cecal carcinoma. Delayed imaging is important if a rebleed or indolent hemorrhage is suspected.[13]
(Reprinted from Winzelberg GG, McKusick KA, Froelich JW, Callahan RJ, Strauss HW. Detection of gastrointestinal bleeding with ^{99m}Tc-labeled red blood cells. *Semin Nucl Med*. 1982;12:139–146. With permission.)

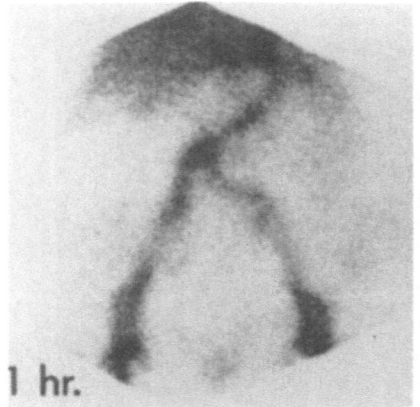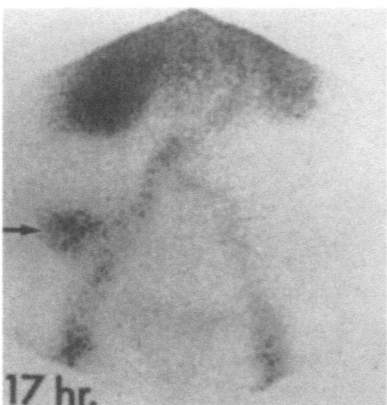

Figure 2.13. Hemorrhage — transverse colon.

Sequential anterior abdominal scintigrams 70 (**A**), 80 (**B**), and 90 (**C**) min after injection. Abnormal activity focus is demonstrated in transverse colon (*arrows*), which appears more intense than hepatic activity. Mean bleeding rate was 0.3 ml/min.[12]
(Reprinted from Smith R, Copley DJ, Bolen FH. 99mTc RBC scintigraphy: correlation of bleeding rates with scintigraphic findings. *AJR*. 1987;148:869–874. With permission.)

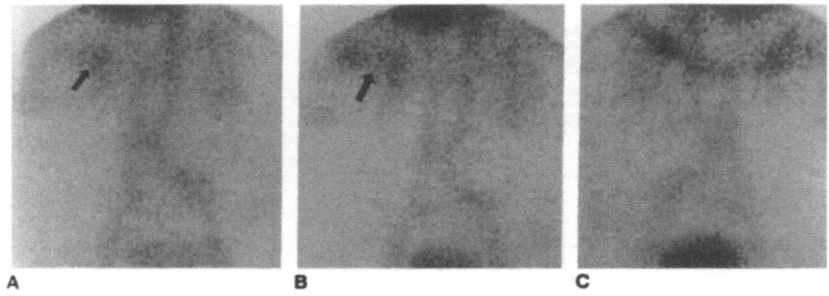

Figure 2.14. Hepatic flexure bleed.

No abnormal activity was shown until 87 min into a routine 90-min study.[14] Therefore, the patient was repositioned and the study extended for another 15 min to illustrate this bleed.
(Reprinted from Bunker SR, Hartshorne MF. Gastrointestinal hemorrhage. In: Mettler, FA ed. *Radionuclide Imaging of the GI Tract*. New York: Churchill Livingstone; 1986: 53–81. With permission.)

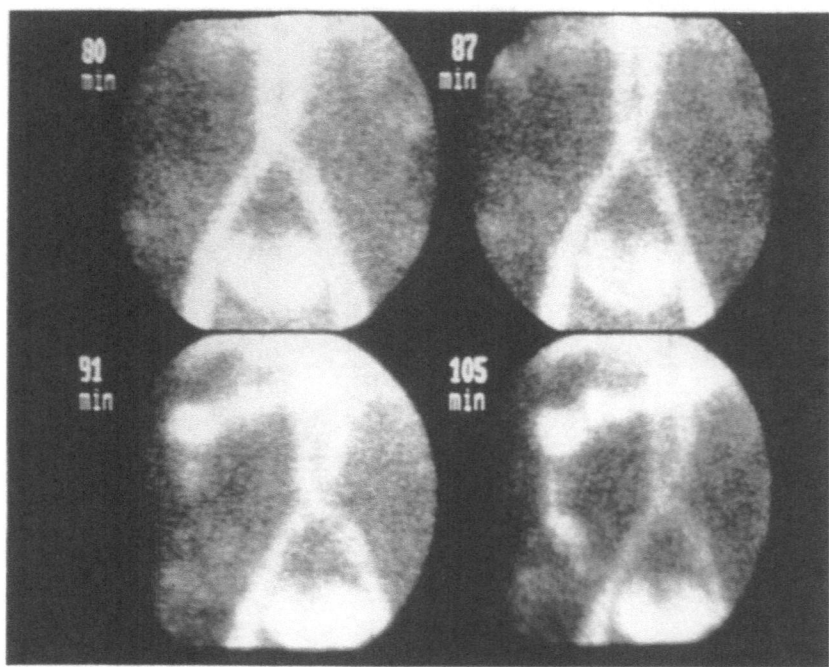

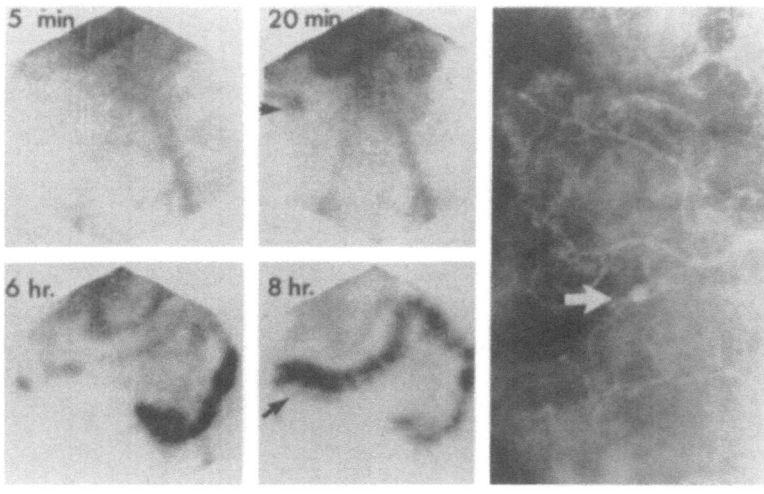

Figure 2.15. Bleeding diverticulum — right colon.

A 78-year-old woman was passing bright red blood per rectum. At 5 min, the scintiscan shows gastric activity but no focal bleeding. At 20 min, focal abnormal uptake is noted in the right colon (*arrow*); however, clinically the patient remained stable. Persistent transfusions were required. At 6 hr, the scintiscan shows marked accumulation of tracer in the transverse and descending colon. At 8 hr there is further bleeding (*arrow*) from the right colon. Angiogram shows a focal area of extravasation from a bleeding diverticulum of the right colon (*arrow*).[19]

(Reprinted from Winzelberg GG, Froelich JW, McKusick KA, et al. Radionuclide localization of lower gastrointestinal hemorrhage. *Radiology*. 1981;139:465–469. With permission.)

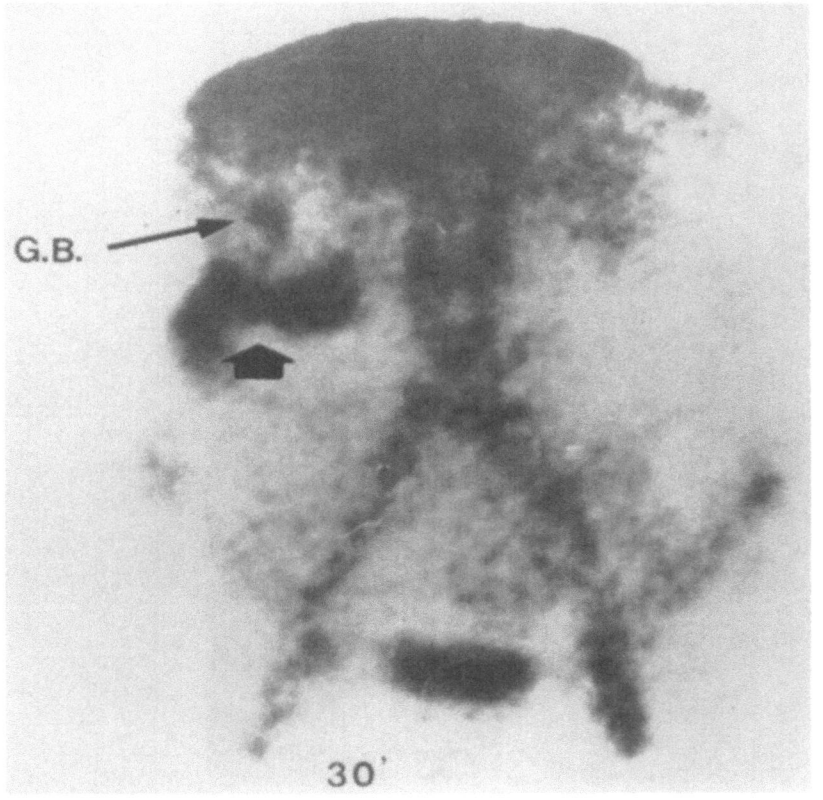

Figure 2.16. Carcinoma — gallbladder.

A gallbladder carcinoma eroding into the hepatic flexure of the colon (*large arrow*). The bleed filled the gallbladder (G.B.) in a retrograde fashion.[20]

(Reprinted from Czerniak A, Zwas ST, Rabau MY, Avigad I, Borag B, Wolfstein I. Scintigraphic demonstration of acute gastrointestinal bleeding caused by gallbladder carcinoma eroding the colon. *Clin Nucl Med*. 1985;10:543–545. With permission.)

Figure 2.17. Hemorrhage — after partial right colectomy.

A $^{99m}TcO_4$ RBC image at 45 min shows lower GI bleeding after a partial colectomy. There is increased activity from the mid-transverse colon through the descending colon (**A**) (*arrowheads*). A superior mesenteric arteriogram demonstrated no extravasation; however, the inferior mesenteric artery injection (**B**) demonstrates the site of bleeding near the anastomosis (*arrows*).[21] (Reprinted from Dorfman GS, Cronan JJ, Staudinger KM. Scintigraphic signs and pitfalls in lower gastrointestinal hemorrhage: the continued necessity of angiography. *RadioGraphics*. 1987;7:543–561. With permission.)

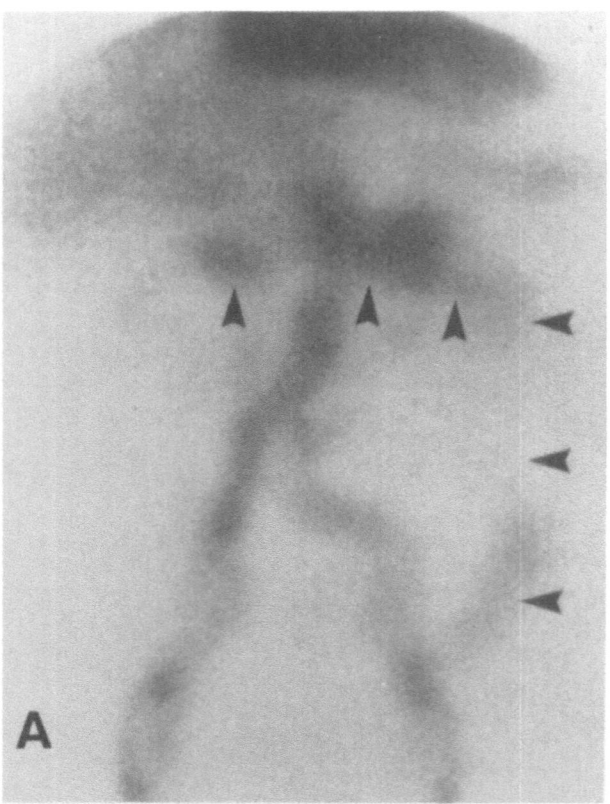

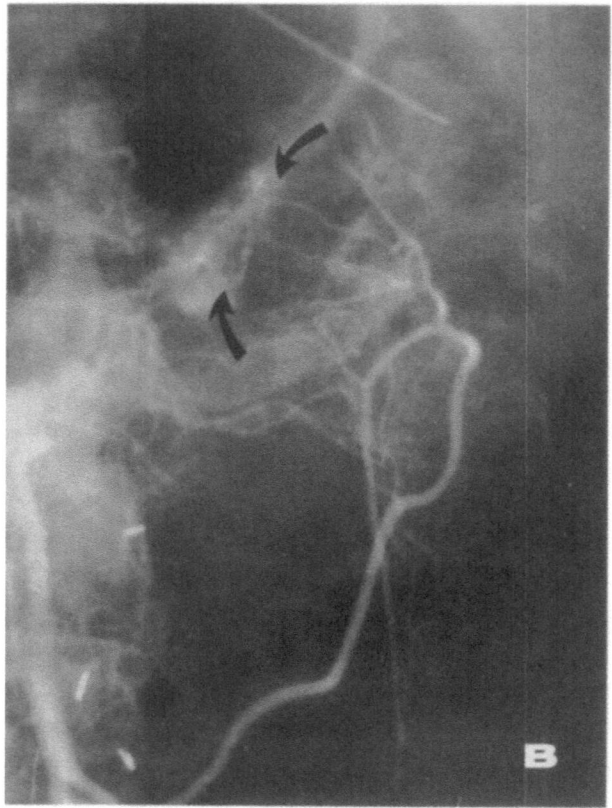

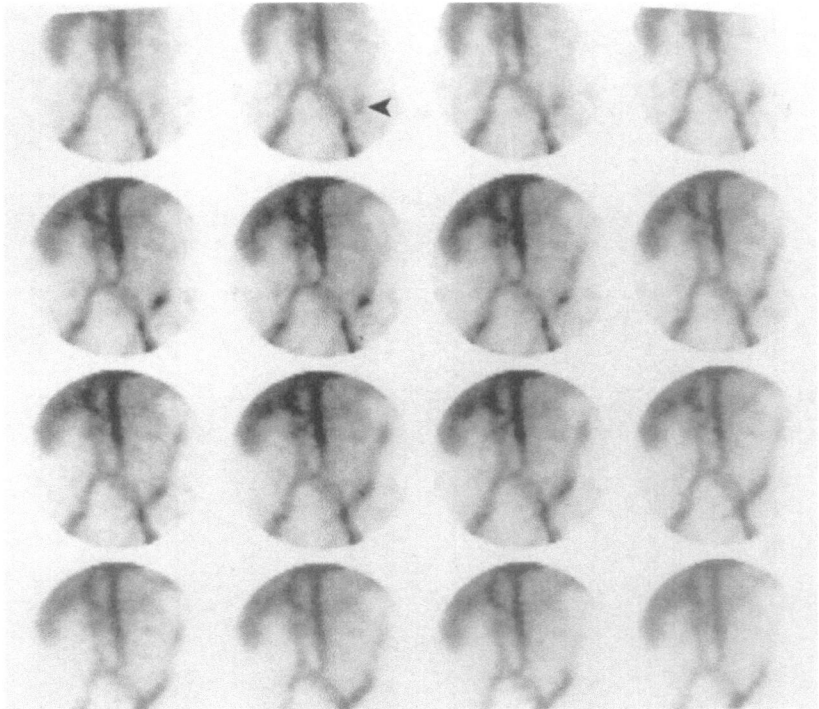

Figure 2.18. Diverticular hemorrhage—colon.

Sequential 1-min images after $^{99m}TcO_4$ RBC injection show a small initial focus of activity (*arrowhead*) accumulating in the descending colon that migrates proximal and distal to the origin as the study progresses. This emphasizes the rapid movement of extravasated blood within the bowel lumen.[14] (Reprinted from Bunker SR, Hartshorne MF. Gastrointestinal hemorrhage. In: Mettler, FA ed. *Radionuclide Imaging of the GI Tract*. New York: Churchill Livingstone; 1986: 53–81. With permission.)

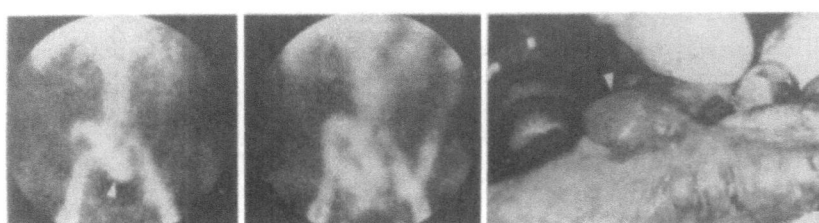

Figure 2.19. Diverticular hemorrhage—sigmoid colon.

Left: 5 min after injection, focal activity is demonstrated in loop of sigmoid colon (*arrowhead*).
Center: image at 25 min confirms sigmoid anatomy and demonstrates rapid movement of radiotracer both distal and proximal to site of origin.
Right: intraoperative photograph shows subserosal extension of diverticular hemorrhage in upper sigmoid loop (*arrowhead*).[22] (Reprinted from Bunker SR, Brown JM, McAuley RJ, et al. Detection of gastrointestinal bleeding sites: use of in vitro technetium Tc 99m-labeled RBCs. *JAMA*. 1982;247:789–792. With permission.)

Figure 2.20. Hemorrhage — sigmoid colon.

A: Three minutes after injection of $^{99m}TcO_4$ RBCs there is focal activity in sigmoid colon (*arrowheads*).

B: By 20 min there is poor localization with passage of radiotracer in stool.

C: Subtraction film selective arteriogram, inferior mesenteric artery injection, shows extravasation from vessels supplying sigmoid colon (*arrowhead*).[22]

(Reprinted from Bunker SR, Brown JM, McAuley RJ, et al. Detection of gastrointestinal bleeding sites: use of in vitro technetium Tc 99m-labeled RBCs. *JAMA*. 1982;247:789–792. With permission.)

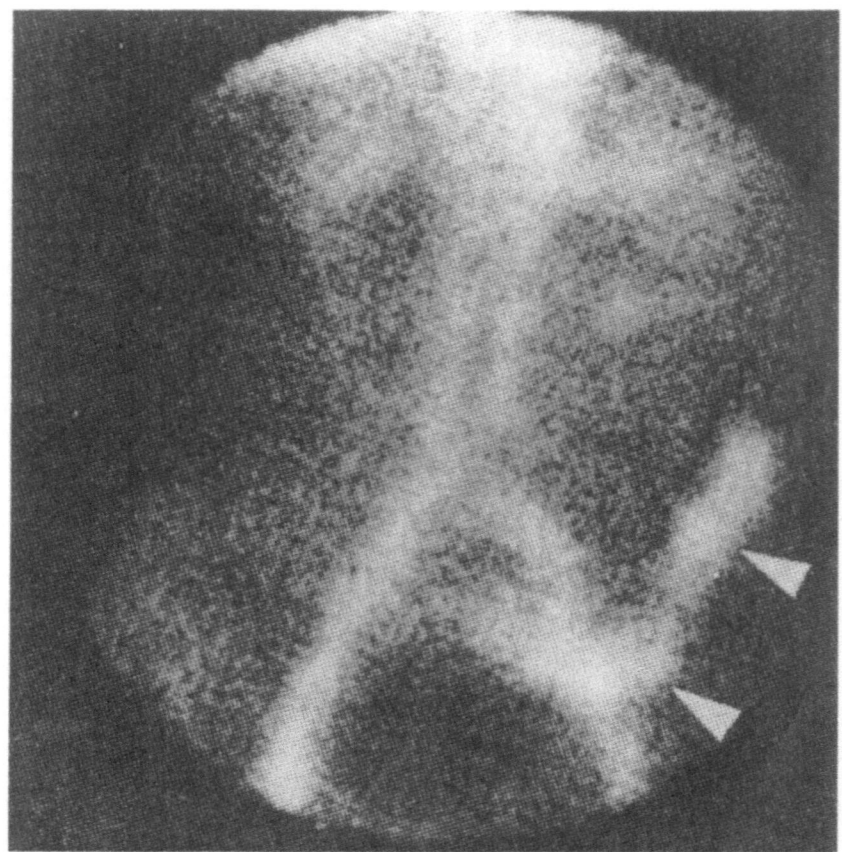

A

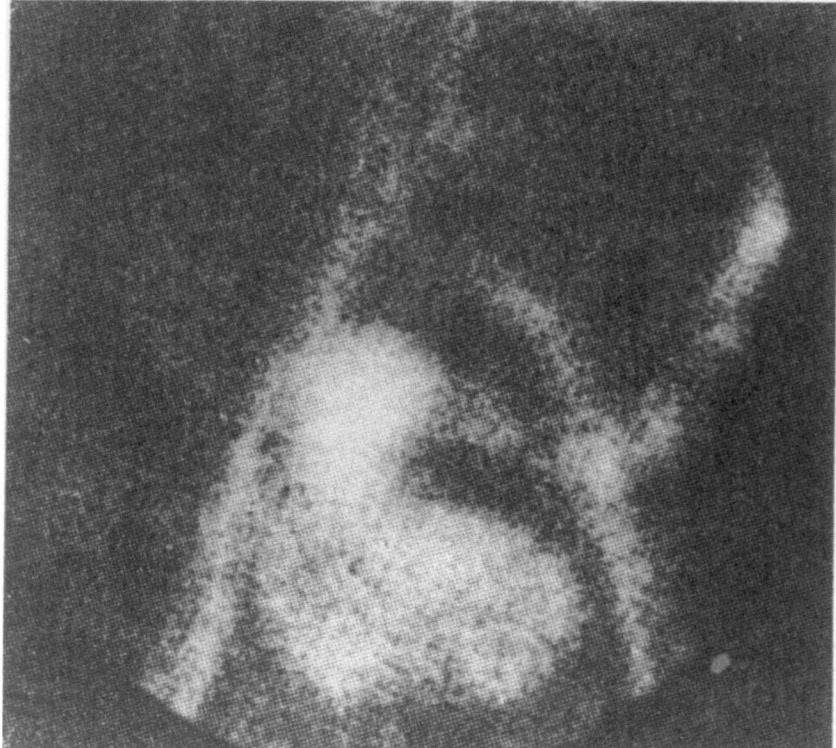

B

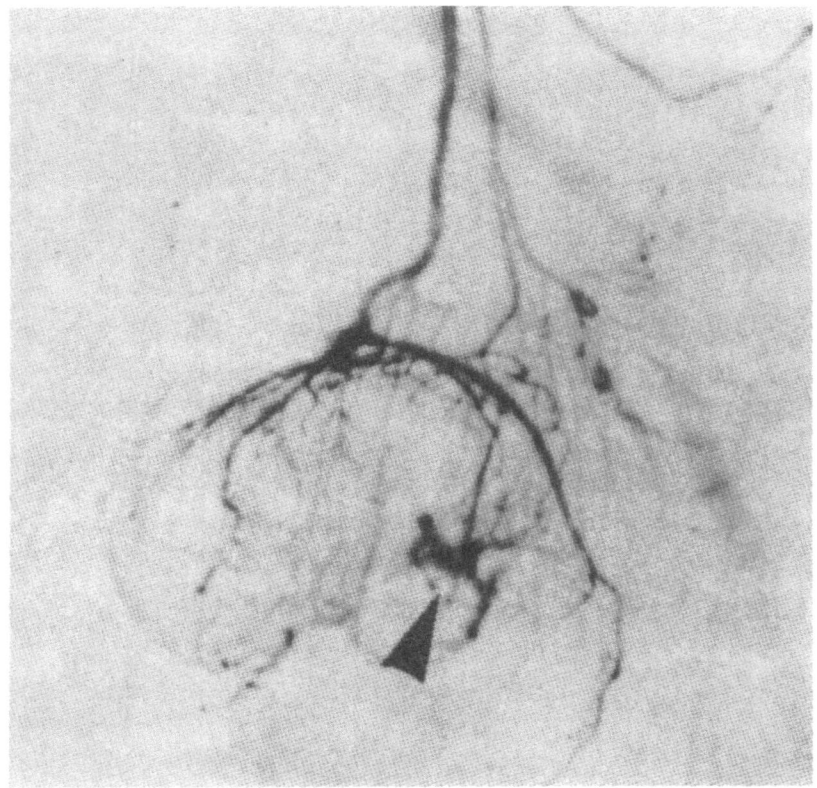

C

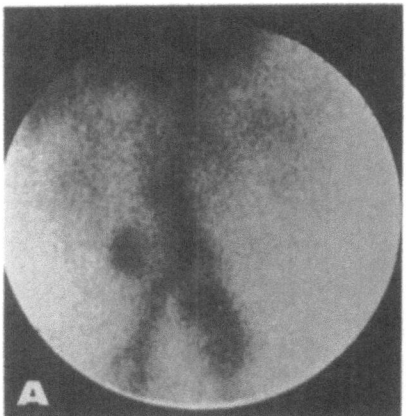 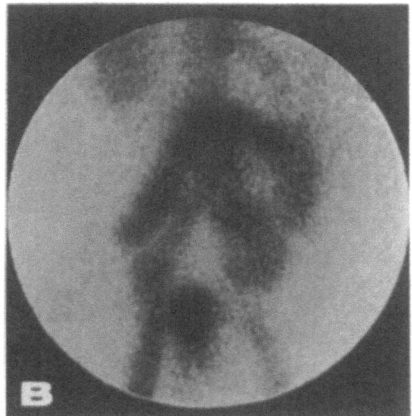

Figure 2.21. Hemorrhage – displaced colon.

Anterior image of the abdomen obtained at 10 min. Presence of a small focus of activity to the right and slightly superior to the aortic bifurcation is demonstrated. Left-sided pelvic kidney is noted to overlie the left iliac vessels. **A:** Subsequent view of the abdomen and pelvis demonstrates spread of the extravasated blood proximally and distally within the lumen of the sigmoid colon. The intraluminal blood clearly delineates the position of the sigmoid colon, which is displaced superiorly and to the right by the left pelvic kidney (**B**).[23]
(Reprinted from Zuckier LS, Patel YD, Kratka PS, Sugarman LA. Pelvic kidney displacing a sigmoid diverticular bleed: scintigraphic and angiographic correlation. *Clin Nucl Med*. 1988;13:463–464. With permission.)

Figure 2.22. Posttraumatic perforation and bleeding — sigmoid colon.

A 56-year-old man presented with bright red blood per rectum after falling and hitting the left side of his abdomen on a table. He had a prior history of surgery of the sigmoid colon. Labeled RBC scintigraphy showed a focus of abnormal activity in the region of the sigmoid colon (*arrow*) seen 25 min after injection (**B**). This was not seen on earlier images (**A**). Images obtained at 35 and 45 min after injection (**C, D**) showed extravascular extravasation along the sigmoid colon into the rectal region. At surgery, a perforated sigmoid colon was found. Emptying bladder and lateral views would be necessary for accurate diagnosis.[24]

(Reprinted from Moreno AJ, Reeves TA, Pearson VD, Rodriguez Turnbull GL. Unusual manifestations of hemorrhage during technetium 99m red cell blood pool imaging. *Clin Nucl Med*. 1989;14:470–471. With permission.)

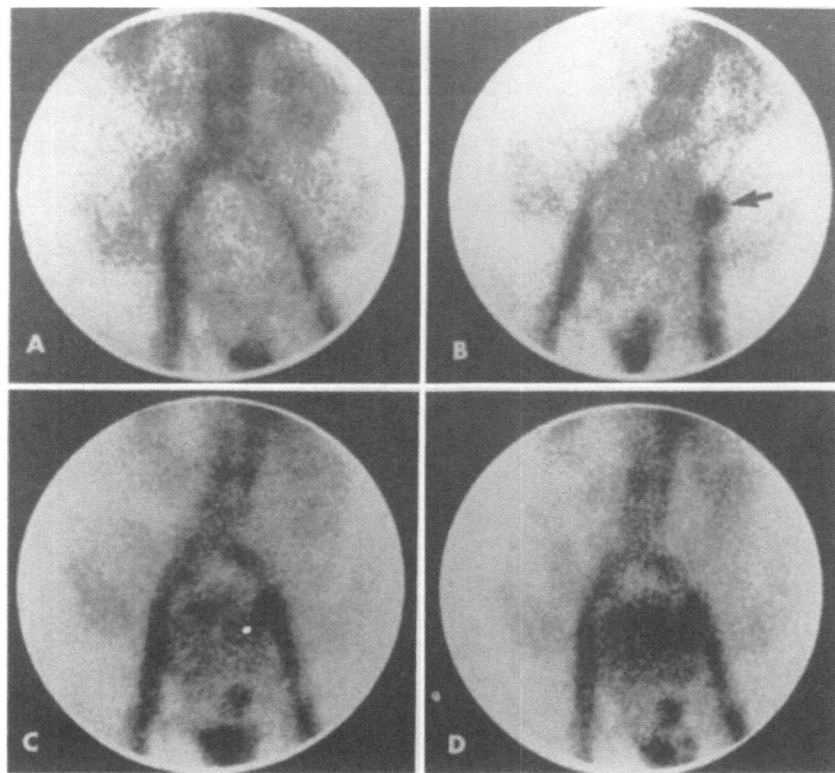

Figure 2.23. Retrograde flow of ⁹⁹ᵐTcO₄ RBC.

Figure 2.23. Retrograde flow of $^{99m}TcO_4$ RBC.

Reverse peristalsis rapidly transfers blood from a sigmoid bleed to the right colon after a bowel movement. Surgery confirmed a single source of hemorrhage in sigmoid colon.[25]

(Reprinted from Wise PA, Sagar VV. Retrograde flow pattern on a gastrointestinal bleeding scan. *Clin Nucl Med*. 1988;13:56. With permission.)

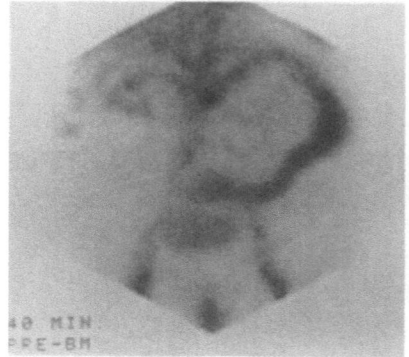 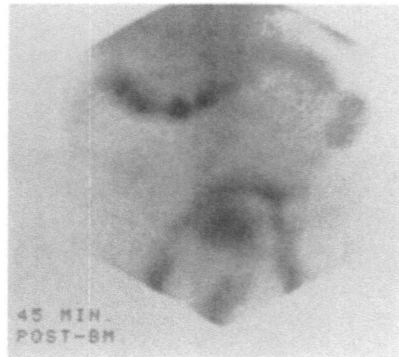

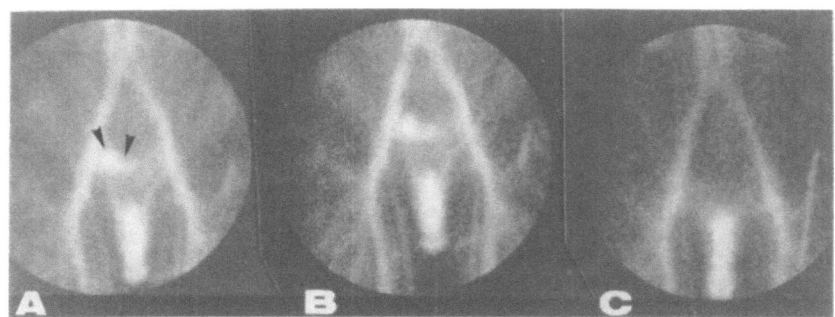

Figure 2.24. Rectosigmoid bleed.

A male patient (**A**) with a small focus of abnormal activity in the pelvis (*arrowheads*). The patient was reimaged (**B**) after voiding urine with no change in the abnormality. After defecation (**C**), the abnormality disappeared.[14]

(Reprinted from Bunker SR, Hartshorne MF. Gastrointestinal hemorrhage. In: Mettler, FA ed. *Radionuclide Imaging of the GI Tract*. New York: Churchill Livingstone; 1986: 53–81. With permission.)

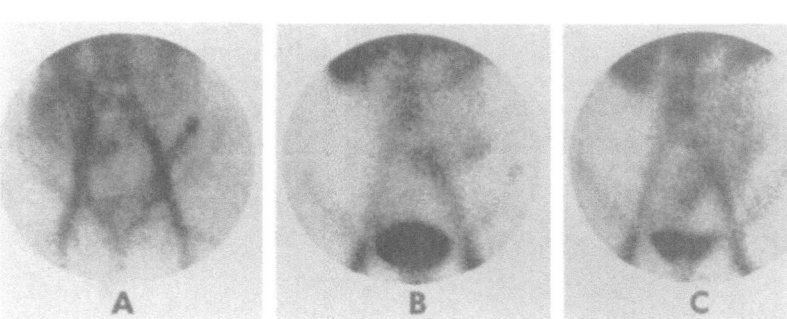

Figure 2.25. Munchausen syndrome.

Anterior view of the abdomen. First examination (**A**) showed radionuclide accumulation in right upper and left lower abdomen. Second examination (**B**) revealed activity in left lower abdomen. Images 1 hr later (**C**) seemed to locate bleeding in descending colon. The explanation is unusual in that the patient had given herself intraabdominal injections of blood that she had drawn from her own arm during the scintigraphic study.[26]

(Reprinted from Bakkers JTN, Crobach LFSJ, Pauwels EKJ. Factitious gastrointestinal bleeding. *J Nucl Med*. 1985;26:667–667. With permission.)

Figure 2.26. Rectal hemorrhage.

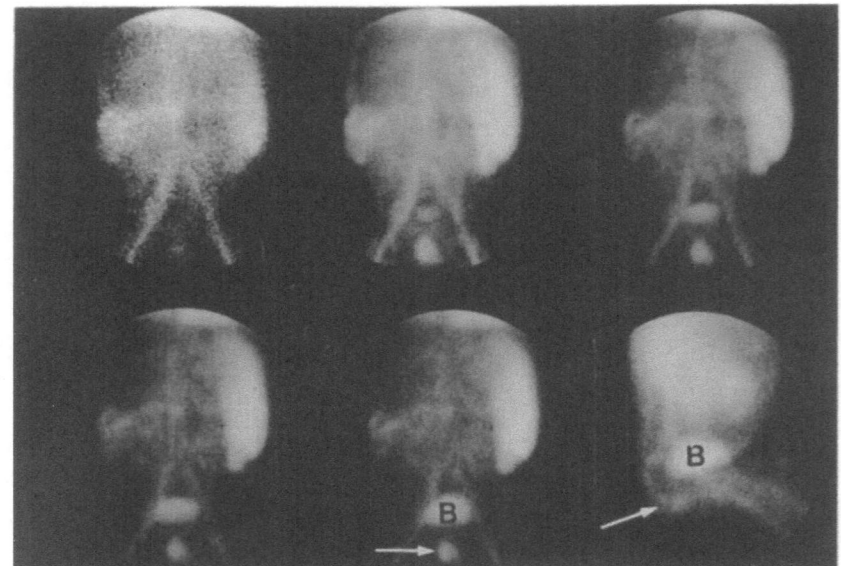

The child had undergone a partially successful Kasai procedure for biliary atresia. A ^{99m}TcO$_4$ RBC (heterologous) study was performed after massive rectal hemorrhage. Upper GI varices had been documented endoscopically but were not bleeding. An immediate image and sequential images over 1 hr show multiple findings that include diminished blood pool activity in the liver, splenomegaly with intense blood pool activity, a hyperemic loop of bowel in the right lateral abdomen (biliary drainage ostomy), and a focus of activity (*arrow*) that appears below the accumulating bladder (B) activity. Note that the abdominal ^{99m}TcO$_4$ RBC activity is shown to be in the rectosigmoid colon (*arrowheads*) on the final right lateral view. Bleeding from a rectal hemorrhoidal vein was later confirmed.[14]

(Reprinted from Bunker SR, Hartshorne MF. Gastrointestinal hemorrhage. In: Mettler, FA ed. *Radionuclide Imaging of the GI Tract*. New York: Churchill Livingstone; 1986: 53–81. With permission.)

Technical Problems and Pitfalls in Interpretation

Figure 2.27. Gastric activity-free pertechnetate.

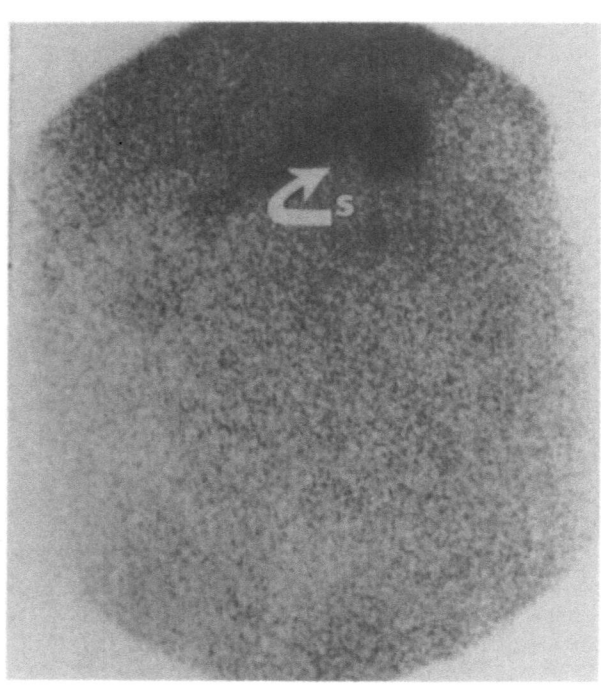

^{99m}TcO$_4$ RBC static image at 40 min shows the result of poor labeling with physiologic ^{99m}TcO$_4$ excretion in the stomach (*curved arrow*). Note the absence of blood pool activity. Imaging of the neck would demonstrate thyroid and salivary gland activity, other signs of free pertechnetate.[21]

(Reprinted from Dorfman GS, Cronan JJ, Staudinger KM. Scintigraphic signs and pitfalls in lower gastrointestinal hemorrhage: the continued necessity of angiography. *RadioGraphics*. 1987;7:543–561. With permission.)

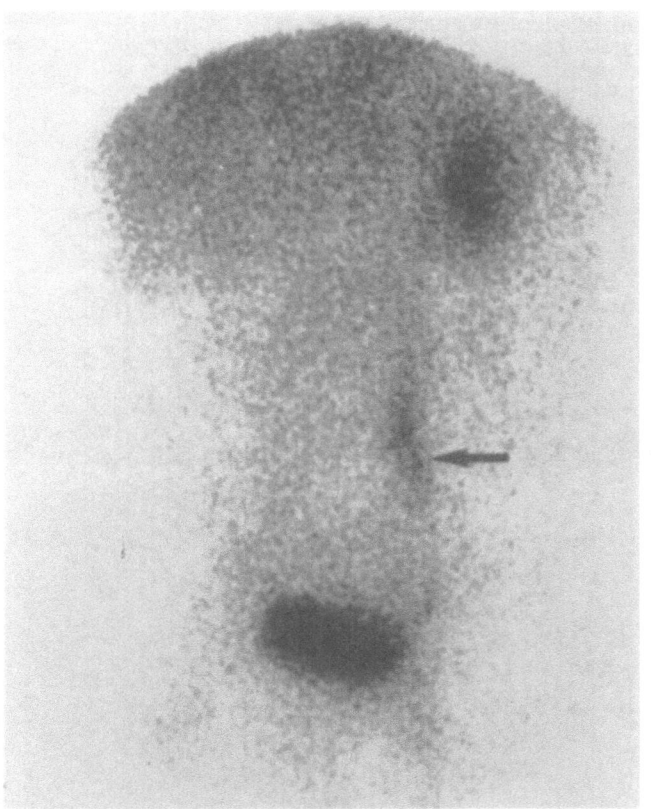

Figure 2.28. Ureter activity.

Activity in the left ureter (*arrow*) at 35 min.[21]
(Reprinted from Dorfman GS, Cronan JJ, Staudinger KM. Scintigraphic signs and pitfalls in lower gastrointestinal hemorrhage: the continued necessity of angiography. *RadioGraphics*. 1987;7:543–561. With permission.)

Figure 2.29. Penile activity.

Anterior image of a ^{99m}TcO$_4$ RBC scan centered over the pelvis (**A**). Note the intense activity in the penis (p, penis). Left (**B**) and right (**C**) anterior oblique images confirm the same normal anatomy.[27]
(Reprinted from Haseman MK. Potential pitfalls in the interpretation of erythrocyte scintigraphy for gastrointestinal hemorrhage. *Clin Nucl Med*. 1982;7:309–310. With permission.)

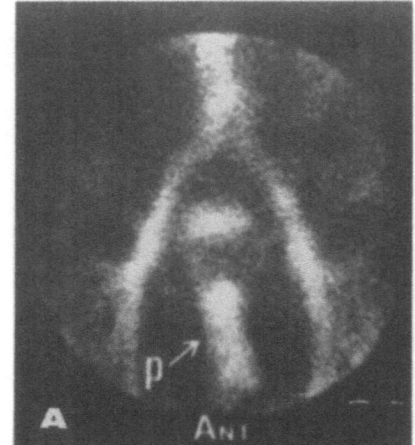
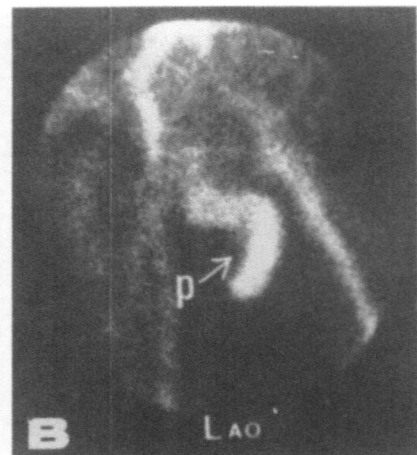
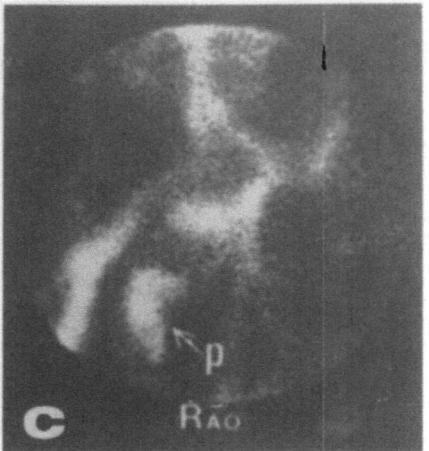

Figure 2.30. Penile activity — erection.

Three sequential images from a ^{99m}TcO$_4$ RBC study showing a focus of activity over the lower pelvis. Note the increase in intensity and change of position. The male patient being examined has experienced penile erection. This particularly intense blood pool with its characteristic form should not be taken for hemorrhage.[14]
(Reprinted from Bunker SR, Hartshorne MF. Gastrointestinal hemorrhage. In: Mettler, FA ed. *Radionuclide Imaging of the GI Tract*. New York: Churchill Livingstone; 1986: 53–81. With permission.)

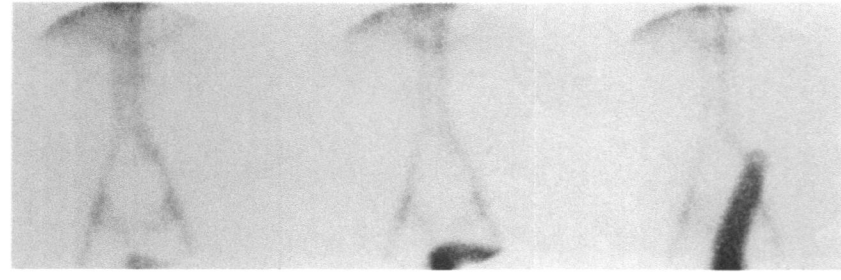

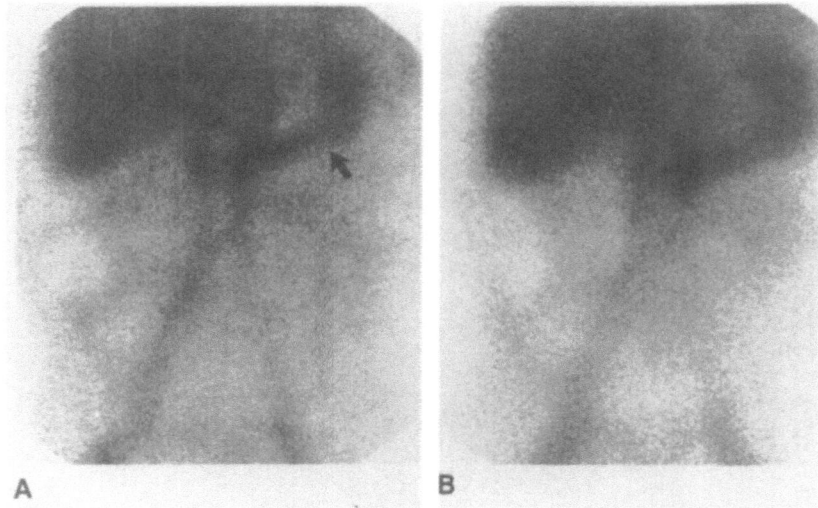

Figure 2.31. Gastritis.

Nearly immediate, moderately intense, and persistent accumulation of ^{99m}TcO$_4$ RBC in stomach on anterior images at 5 (**A**) and 45 (**B**) min postinjection. Multiple gastric erosions were found endoscopically. No active bleeding was identified to 60 min, as evidenced by lack of advancement of activity out of stomach. No renal excretion was seen to suggest presence of free ^{99m}TcO$_4$ in this patient with gastritis.[28]
(Reprinted from Wilton GP, Wahl RL, Juni JE, Froelich JW. Detection of gastritis by ^{99m}Tc-labeled red-blood-cell scintigraphy. *AJR*. 1984;143:759–760. With permission.)

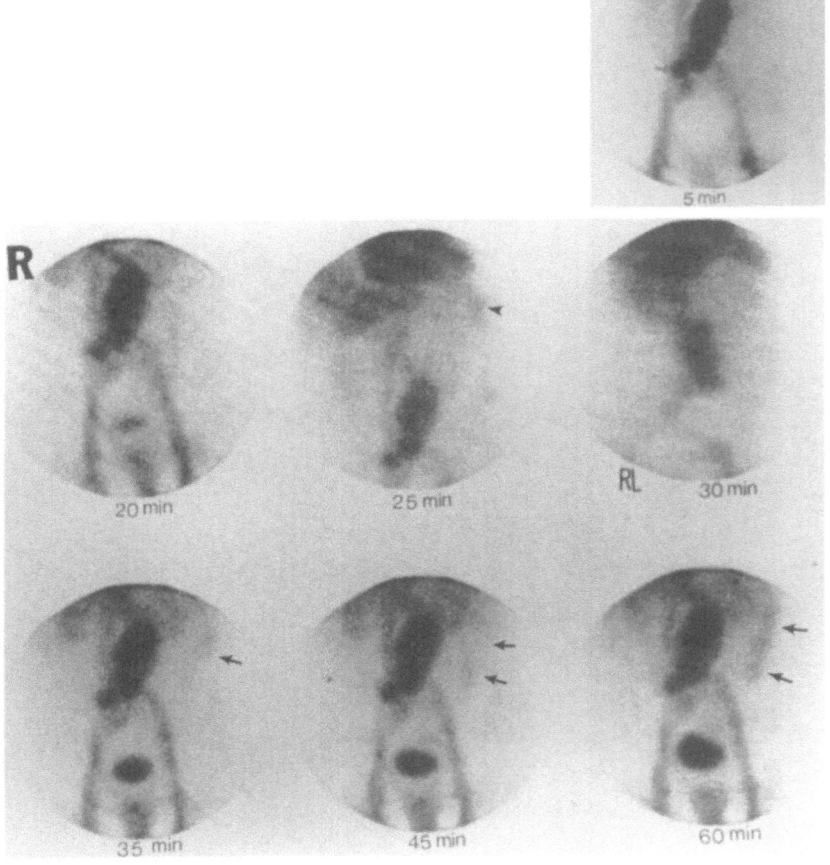

Figure 2.32. Incidental abdominal aneurysm unrelated to active bleed.

Abnormal radiotracer localization is seen in the abdomen as indicated by the arrow heads (5 min), and is more apparent in the splenic flexure (*arrowheads*, 25 min) and descending colon (*arrows*) in the 60-min images. A large abdominal aneurysm with extension to the proximal iliac artery on the right (*open arrow*) is persistently visualized throughout the study. Cardiomegaly is evident in the 25-min image. Bleeding adenocarcinoma of the transverse colon accounted for the bowel activity.[29]
(Reprinted from Shih WJ, Magoun S, Donstad PA. Huge abdominal aortic aneurysm demonstrated by technetium 99m labeled RBC blood pool imaging. *Clin Nucl Med*. 1987;12:825–826. With permission.)

Figure 2.33. Abdominal aorta.

An ectatic abdominal aorta (Ao) is seen on a 40-min ^{99m}TcO$_4$ RBC study, and the radioactivity seen does not meet the criteria for positive GI hemorrhage (see Table 2.4).[21]

(Reprinted from Dorfman GS, Cronan JJ, Staudinger KM. Scintigraphic signs and pitfalls in lower gastrointestinal hemorrhage: the continued necessity of angiography. *RadioGraphics*. 1987;7:543–561. With permission.)

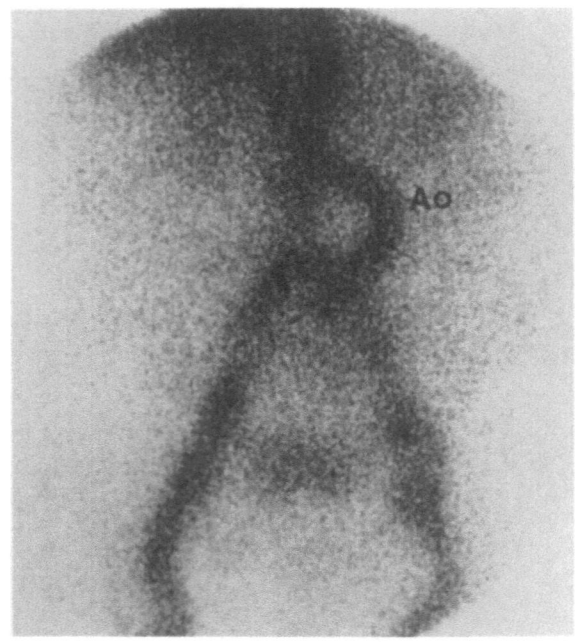

Figure 2.34. Pseudoaneurysm.

Scintigrams at 10 min, 1 and 14 hr postinjection show a round tracer accumulation outside the vascular bed in a patient with a false aneurysm of the thigh. P is the external marking of the site where the inguinal pulse of a the femoral artery was palpable. Confirmed by distal subtraction angiography.[30]

(Reprinted from Fritsch S, Arleth J, Becker HM, Mannes AG, Moser E. Detection of a false aneurysm in the thigh by scintiangiography: case report. *Clin Nucl Med*. 1987; 12:564–566. With permission.)

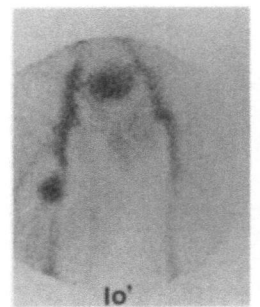 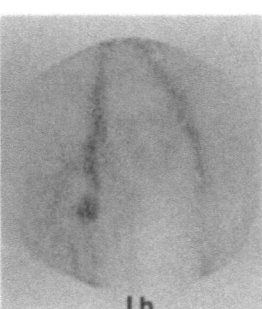 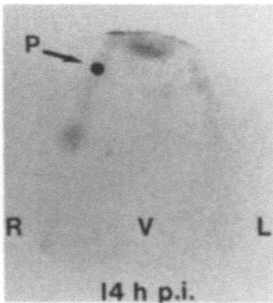

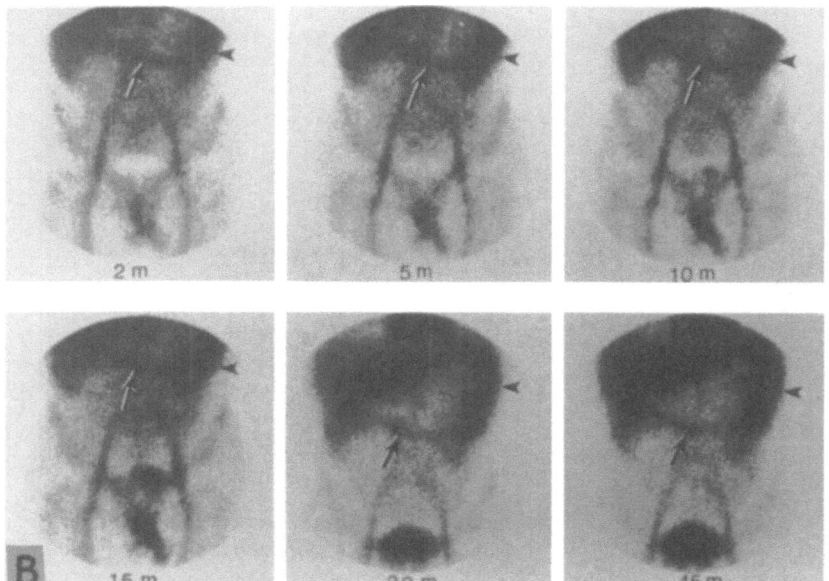

Figure 2.35. Gastric varices.

Images at 2, 5, 10, 15, 30, and 45 min showing persistent activity located in the gastric area (*arrowheads*) and the midabdominal area (*arrow*), which proved to be gastric varices.[31]

(Reprinted from Shih WJ, Domstad PA, Loh FG, Pulmano C. Extensive gastric varices demonstrated by technetium 99m red blood cell scintigraphy. *Clin Nucl Med*. 1987;12:290–293. With permission.)

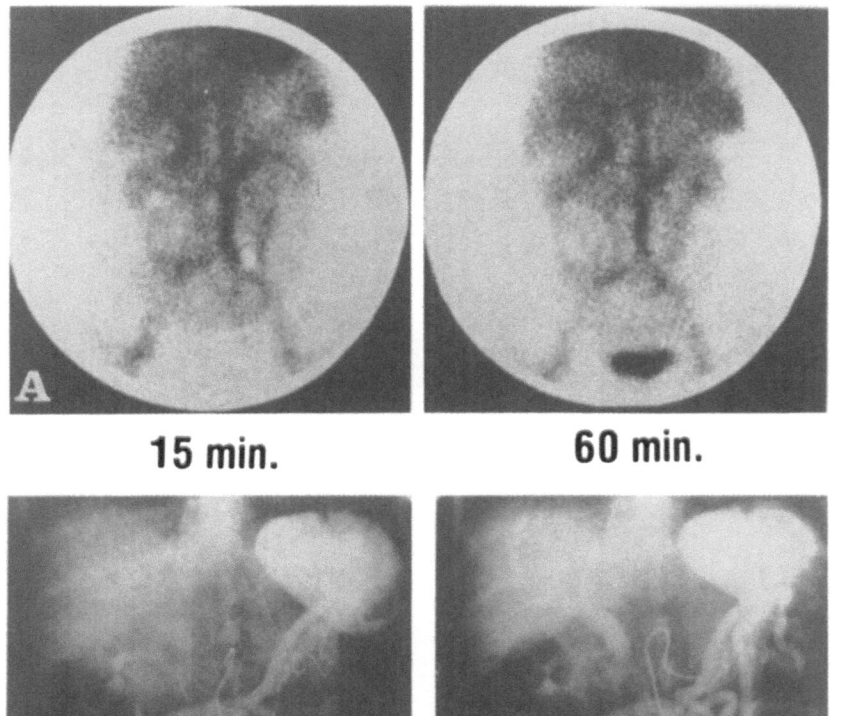

Figure 2.36. Portal venous collaterals.

Images at 15 and 60 min demonstrate persistent changes in the same distribution. Films from the venous phase of the superior mesenteric and celiac arteriograms demonstrate the portal and splenic veins to be occluded at the confluence with collateral reconstitution of the portal vein corresponding to the abnormalities seen on the bleeding study.[32]

(Reprinted from Iles S, Gordon D, Aquino JA, Caines J. Visualization of portal venous collaterals on a technetium 99m-labeled red blood cell scan for gastrointestinal bleeding: potential pitfall in interpretation. *Clin Nucl Med*. 1987;9:946–948. With permission.)

Figure 2.37. Transverse mesocolon varix.

Images from a ^{99m}TcO$_4$ RBC study at 5 (**A**), 10 (**B**), 35 (**C**), and 45 (**D**) min has abnormal activity at the splenic flexure that gradually involves the transverse colon (*straight arrows*) and small bowel (*arrowheads*). Early dynamic images were unremarkable and later dynamic images were virtually identical to the 5-min static image. A superior mesenteric angiogram was performed, the venous phase of which reveals a large varix (*arrowheads*) in the transverse mesocolon (**E**). This simulated colonic bleeding by gradually filling with tagged RBCs. The patient was bleeding from hemorrhoids.[21] (Reprinted from Dorfman GS, Cronan JJ, Staudinger KM. Scintigraphic signs and pitfalls in lower gastrointestinal hemorrhage: the continued necessity of angiography. *RadioGraphics*. 1987;7:543–561. With permission.)

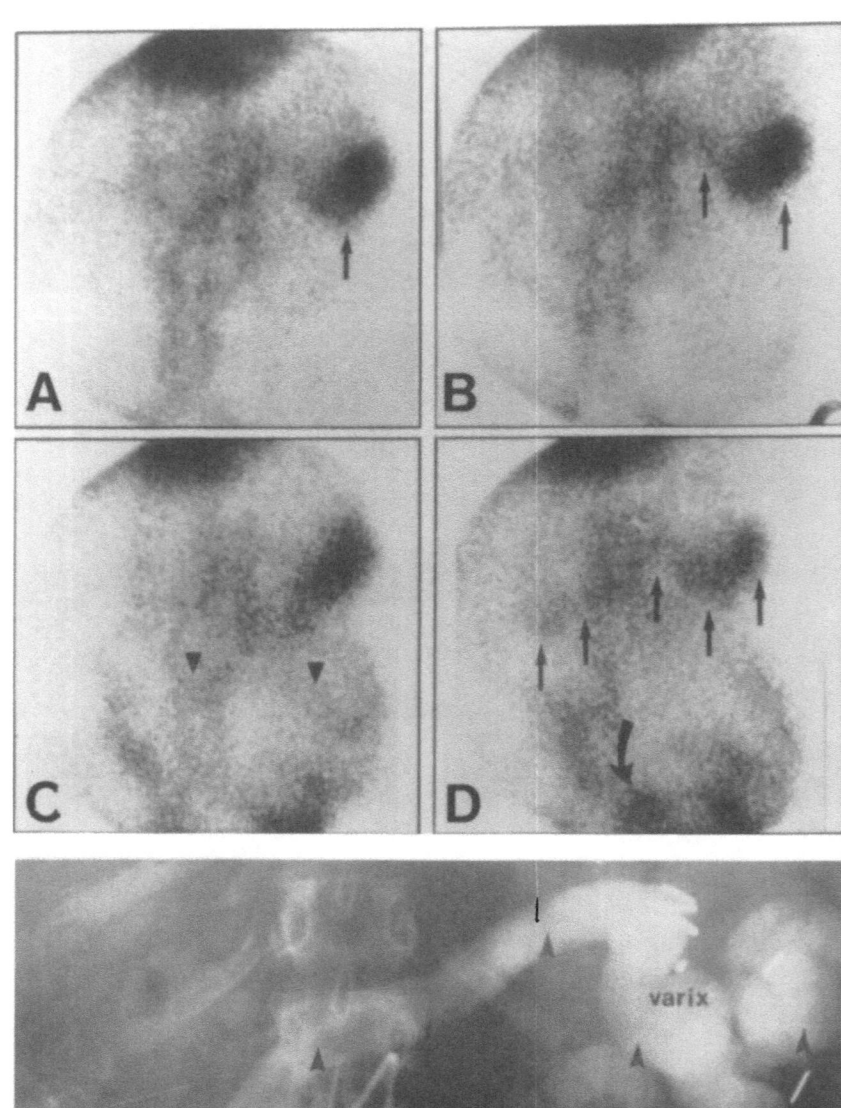

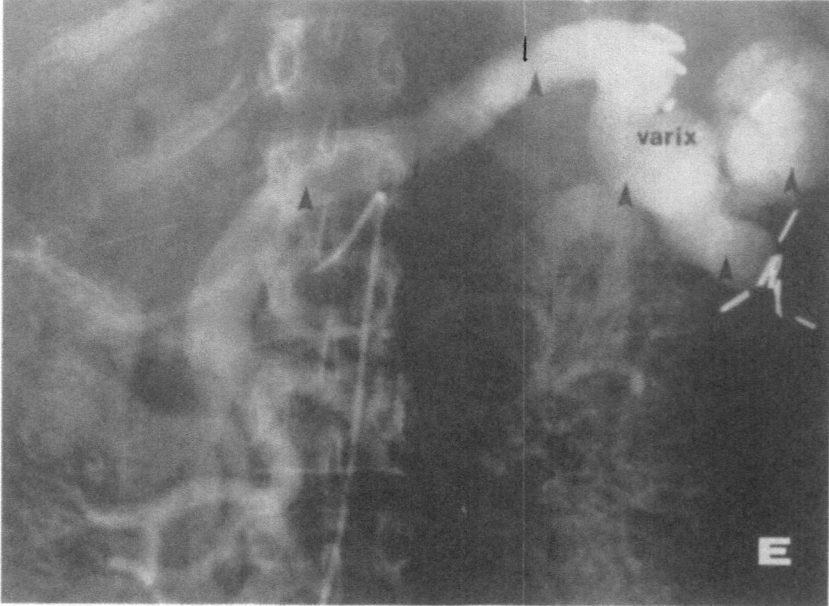

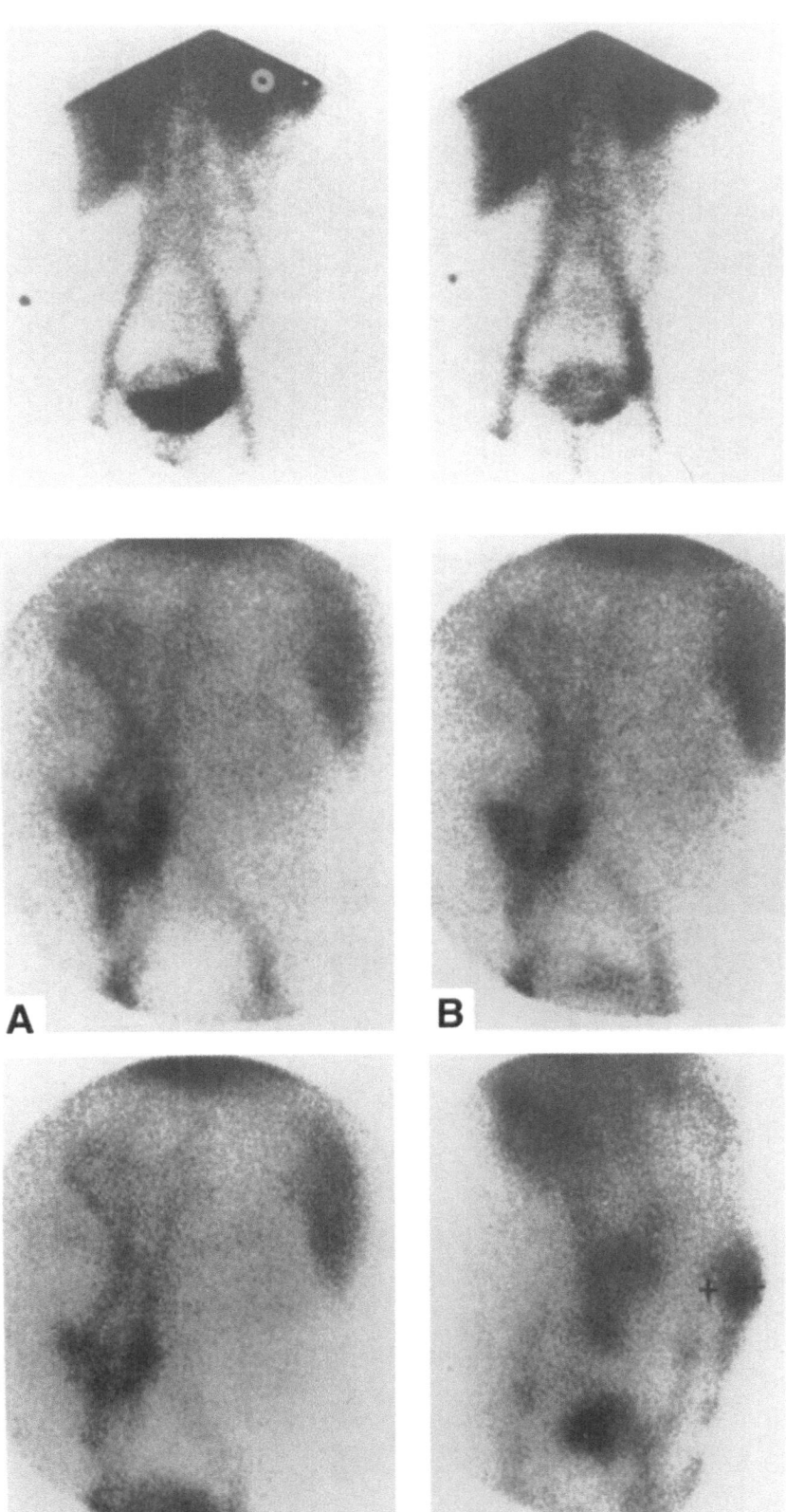

Figure 2.38. Ovarian vein.

Prevoid (*left*) and 60 min postvoid (*right*) images reveal persistence of linear activity with extension into the left upper quadrant. This was found to be a large left ovarian vein at angiography.[33]
(Reprinted from Camele RA, Bansal SK, Turbiner EH. Red blood cell gastrointestinal bleeding scintigraphy: appearance of the left ovarian vein. *Clin Nucl Med*. 1984; 9:275–276. With permission.)

Figure 2.39. Varices—abdomen.

The immediate blood pool image (**A**) and the 45-min postinjection (**B**) reveal the same midabdominal region of increased radiotracer uptake seen on the dynamic study with additional localization extending linearly in a cephalad direction. No changes in size, intensity, or location of the radiotracer uptake is noted in the 45-min image. The abnormal radiotracer activity is again noted unchanged in the 85-min image (**C**). A lateral abdominal image (**D**) at 90 min postinjection shows a part of the increased radiotracer uptake to be located far anteriorly in the region of the caput medusae (*crosses*). Additional activity is seen in dilated intraperitoneal collateral veins.[34]
(Reprinted from Moreno AJ, Byrd BF, Berger DE, Turnbull GL. Abdominal varices mimicking an acute gastrointestinal hemorrhage during technetium 99m red blood cell scintigraphy. *Clin Nucl Med*. 1985;10:248–251. With permission.)

Figure 2.40. Duplicated inferior vena cava.

Delayed images demonstrate vascular channels in the right and left paraspinal space (*arrowheads*). Between these, a third vessel curving up to the right is noted (*arrow*). This is a duplication of the inferior vena cava.[35]

(Reprinted from Howard JL, Dhekne RD, Moore WH, Long SE. Double inferior vena cava by technetium 99m labeled RBC study. *Clin Nucl Med*. 1988;13:671–672. With permission.)

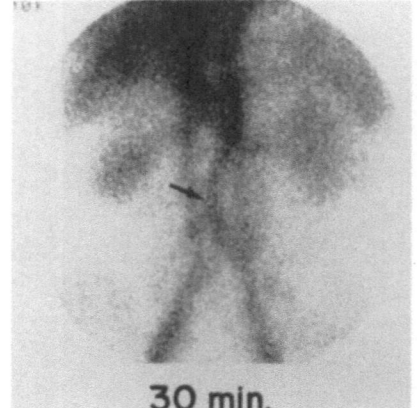
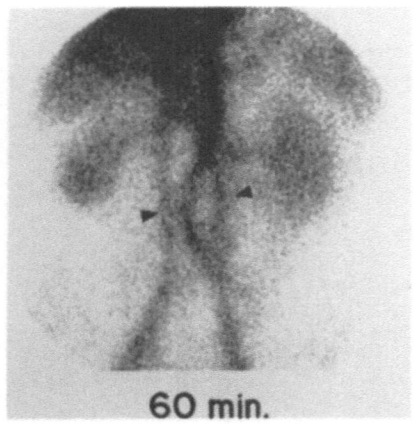

Figure 2.41. Iliac cavernous hemangioma.

Anterior images of the abdomen obtained at 10 mins (**A**), 30 mins (**B**), and (**C**) 16 hr after the injection of $^{99m}TcO_4$ RBCs. Note the accumulation of a well defined area of abnormal activity in the lower abdominal region (*arrow*). The area of abdominal activity remains localized throughout the study with no migration to suggest active bleeding. Surgery confirmed a cavernous hemangioma in the proximal ileum.[36]

(Reprinted from Holoway H, Johnson J, Sandler M. Detection of an ileal cavernous hemangioma by technetium 99m red blood cell imaging. *Clin Nucl Med*. 1988;13:32–35. With permission.)

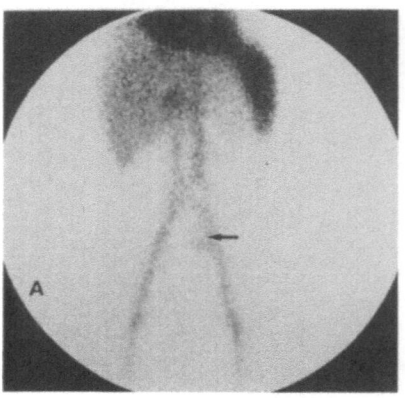
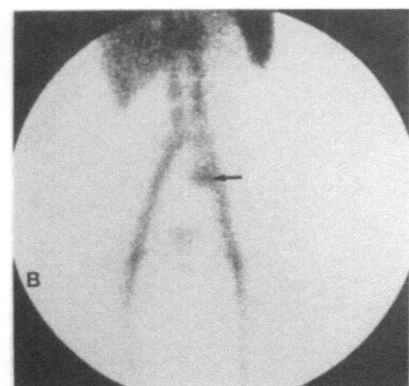
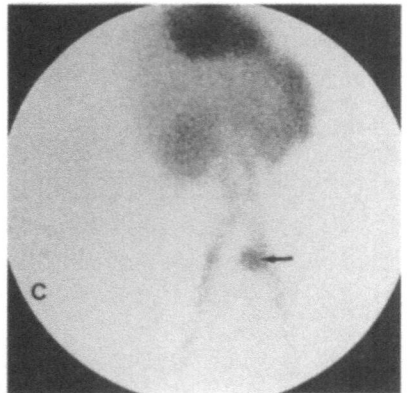

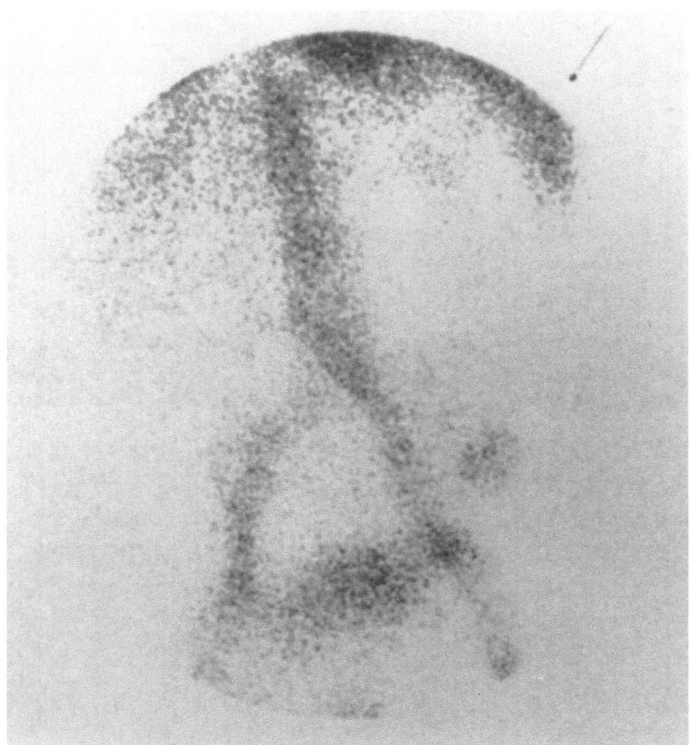

Figure 2.42. Small bowel leiomyoma.

Area of unchanging blood pool is identified in the RLQ superior to the iliac artery. Surgery found a distal ileal leiomyoma that on histology demonstrated numerous dilated vascular channels.[37]
(Reprinted from McDonald K. Technetium 99m RBC scintigraphy in the evaluation of small bowel leiomyoma. *Clin Nucl Med.* 1987;12:131–133. With permission.)

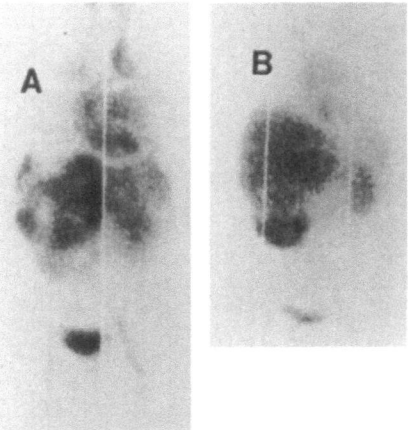

Figure 2.43. Hepatic hemangioma.

Progressive mixing of $^{99m}TcO_4$ RBCs in a huge cavernous hemangioma of the liver in a 65-year-old woman. Early whole body blood pool study in the anterior view (**A**) shows incomplete mixing with areas showing no uptake of the labeled red cells. Late study (**B**) homogeneous filling of the entire lesion. $^{99m}TcO_4$ RBC liver scans are discussed in a separate chapter in this atlas.[38]
(Reprinted from Front D, Ora I, Groshar, Weininger J. Technetium 99m-labeled red blood cell imaging. *Semin Nucl Med.* 1984; 14:226–250. With permission.)

Figure 2.44. Hematobilia.

Anterior (*left*) and lateral (*right*) views at 24 hr show the radionuclide accumulating in the gallbladder. Uremia and hemodialysis are associated with autohemolysis. The hemoglobin from the hemolysis is phagocytized by the Kupffer cells and the porphyrin portion converted to bilirubin, which is subsequently removed by the polygonal cells and excreted in bile. This suggests that the radionuclide may be bound to the porphyrin portion of the hemoglobin molecule or this may represent hematobilia.[39]
(Reprinted from Wood MJ, Hennigan DB. Radionuclide tagged red blood cells in the gallbladder. *Clin Nucl Med*. 1984;9:289–290. With permission.)

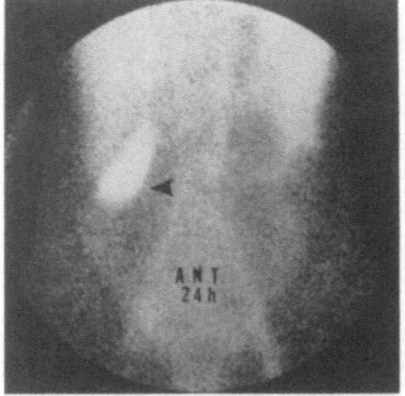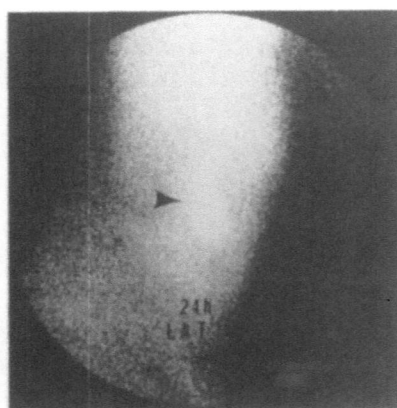

Figure 2.45. Hematobilia.

An image taken 24 hr after the administration of $^{99m}TcO_4$ RBCs demonstrates intense activity in the dilated biliary tract (*arrow*) and less activity in the bowel. This was confirmed on cholangiogram (not shown). Note that the contour of the biliary system as outlined by the labeled RBCs conforms closely to the main biliary ducts. Patient had melanotic stools secondary to biliary hemorrhage from percutaneous stone removal.[40]
(Reprinted from Lee SM, Lee RG, Clouse ME, Hill TC. Demonstration of hematobilia using technetium 99m labeled red blood cells. *Clin Nucl Med*. 1986;11:52. With permission.)

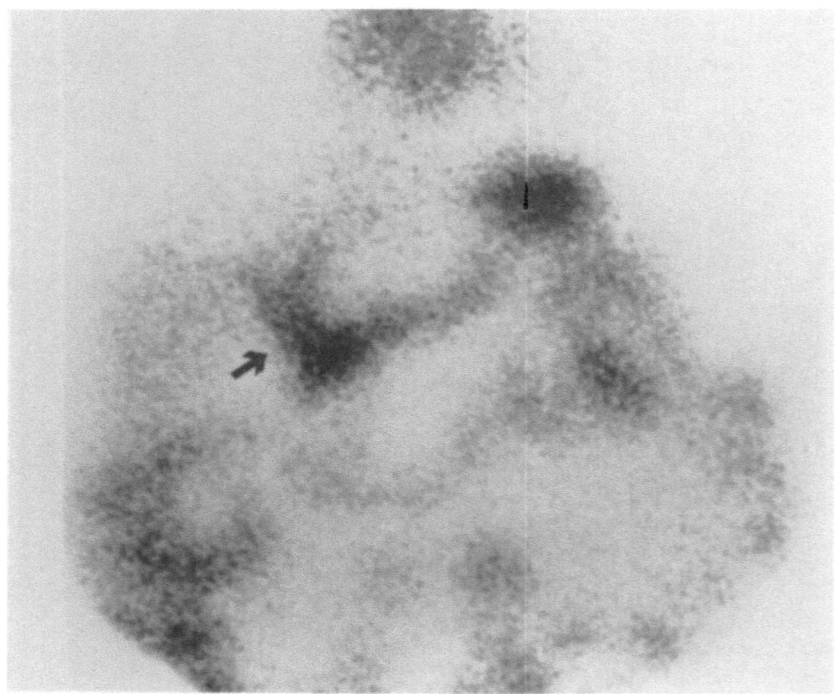

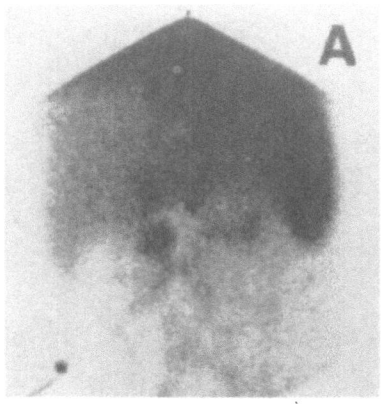
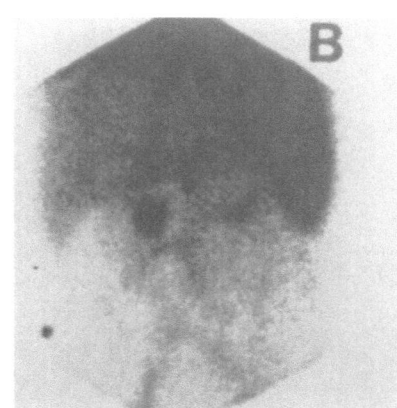
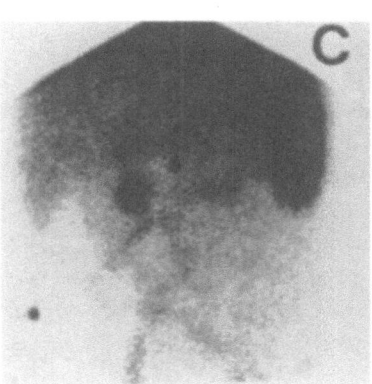
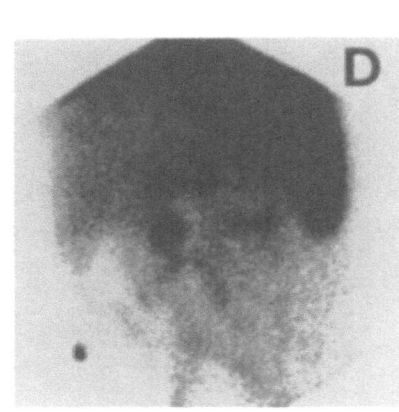

Figure 2.46. Pancreatic pseudocyst.

A $^{99m}TcO_4$ RBC scan confirms hemorrhage within a known pancreatic pseudocyst. Serial images taken every 3 min for 60 min. The persistent linear foci in the LLQ and RLQ represent a splenorenal shunt and a possible varix, respectively, as shown at prior angiography. The image at 1 min (**A**) shows a round area of accumulation of labeled RBCs in the region of the head of the pancreas. This focus increases in intensify (**B** at 5 min) over time and persists throughout the study (**C** at 20 min and **D** at 60 min).[41]

(Reprinted from Ellison MJ, Thornburg A, Turbiner E. Demonstration of bleeding into a pancreatic pseudocyst on a technetium 99m labeled red blood cell scan. *Clin Nucl Med*. 1987;12:969. With permission.)

Figure 2.47. Pediatric retroperitoneal hemorrhage and duodenal hematoma.

A 10-month-old male infant was admitted to the intensive care unit because of multiple trauma, possibly secondary to abuse. There was a faint bluish discoloration over the right flank and midabdomen and blood was obtained from the nasogastric tube and per rectum. Laboratory studies revealed elevated serum amylase and slightly decreased serum calcium levels. An abdominal ultrasound on the second day revealed generalized enlargement of the pancreas but was otherwise unremarkable. Esophagastroduodenoscopy on the fourth day revealed an obstructing blood clot in the pylorus. Images 1.5 hr after the injection of $^{99m}TcO_4$ RBCs demonstrate pooling of activity in the region of the duodenum on both the anterior (**A**) and right lateral (**B**) views (*arrows*). The rectangular photopenic marker on the anterior view identifies the umbilicus. Diagnostic considerations included duodenal hematoma, retroperitoneal hemorrhage, and small bowel hemangioma. An anterior view at 3.5 hr shows an increase in intensity of the upper abdominal activity with no identifiable activity in the distal small bowel or colon (**C**). There was no change in location on the lateral view. An image at 20 hr postinjection demonstrates persistence of the activity without progression of the isotope into distal portions of the gastrointestinal tract (**D**). These findings were believed to exclude an active GI bleed. UGI and ultrasound demonstrated findings compatible with a retroperitoneal hemorrhage and duodenal hematoma.[42]

(Reprinted from Mitchell DS, Stacy TM, Grunow JE, Leonard JC. Scintigraphic detection of an occult abdominal bleed in a child. *Clin Nucl Med*. 1988;13:546–548. With permission.)

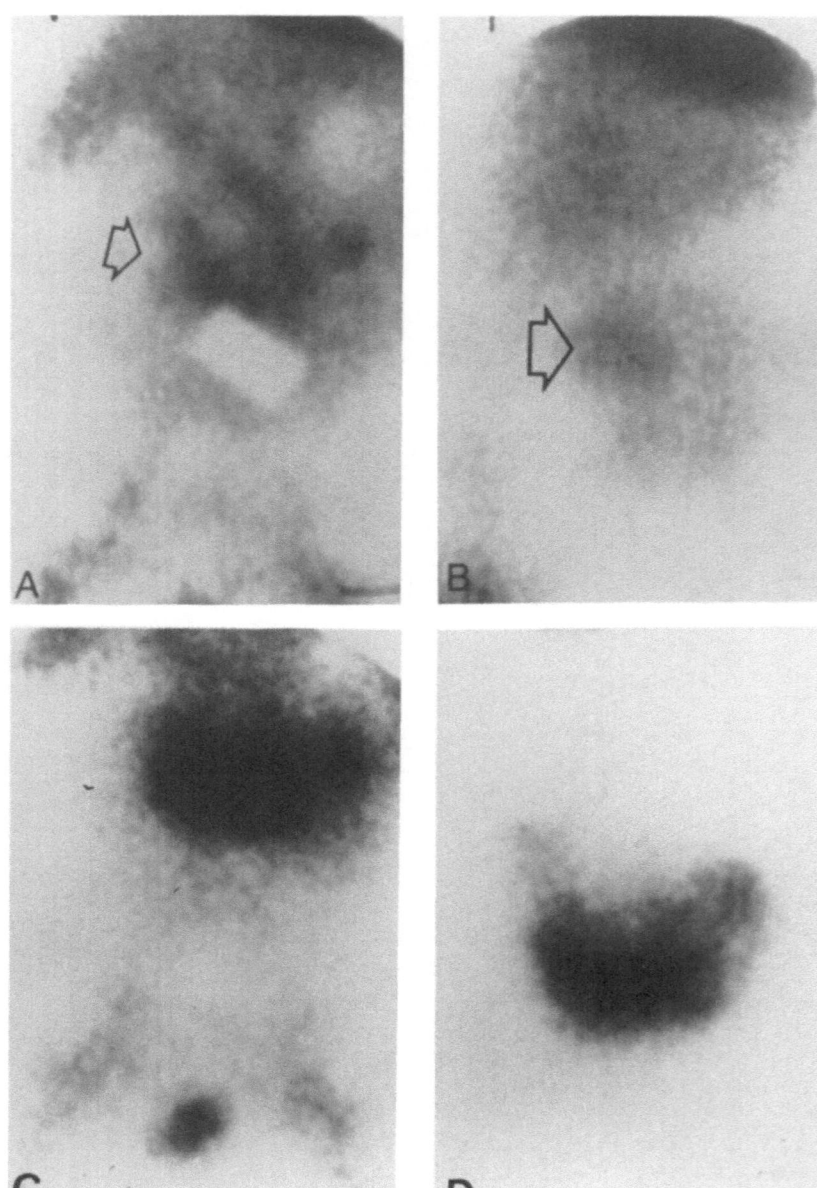

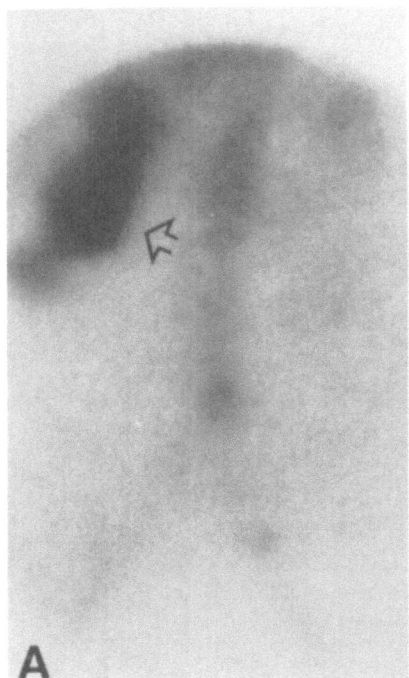

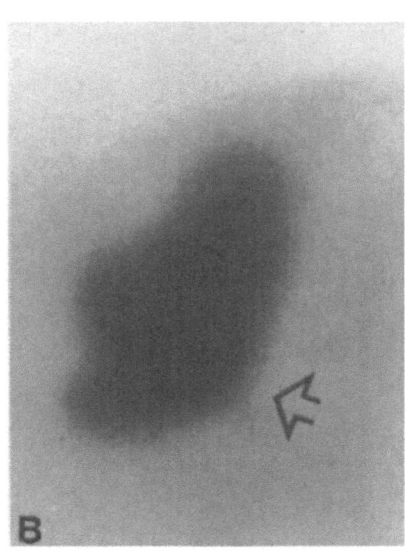

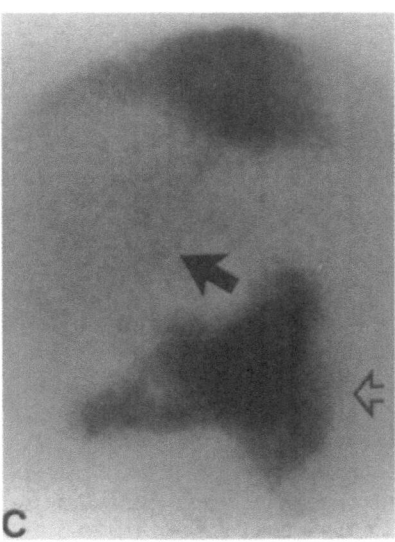

Figure 2.48. Hemorrhage—Morrison's pouch.

Anterior projection (**A**) shows collection of tagged RBCs 20 min postinjection (*open arrow*). Sixty minutes postinjection (**B**) spot image shows no change. RAO projection (**C**) shows liver (*solid arrow*) and amorphous pooling of tagged RBCs (*open arrow*). This was proven by ultrasound to be extravasated $^{99m}TcO_4$ RBCs into Morrison's pouch. Surgery demonstrated a bleeding hepatocellular carcinoma.[43]
(Reprinted from Czarnecki DJ. Intraperitoneal hemorrhage diagnosed by technetium 99m labeled RBC imaging. *Clin Nucl Med*. 1986;11:617–618. With permission.)

Figure 2.49. Gluteal hemorrhage.

A 24-hr posterior $^{99m}TcO_4$ RBC lower abdominal and pelvic scintiscan confirms the presence of an abnormal accumulation of radiotracer within the region of the left superior iliac crest (*arrow*). This represented an incidental finding in a study performed for guaiac positive stools.[44]
(Reprinted from Fink-Bennett D, Johnson JR. Gluteal hematoma: a potential cause for a false-positive Tc-99m RBC gastrointestinal bleeding study. *Clin Nucl Med.* 1984;9:414. With permission.)

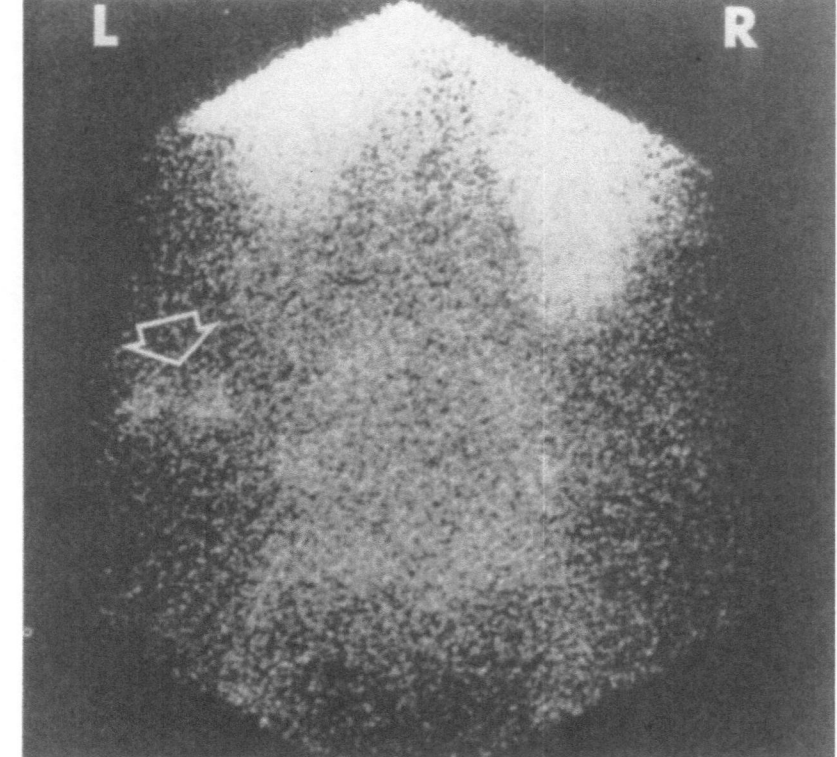

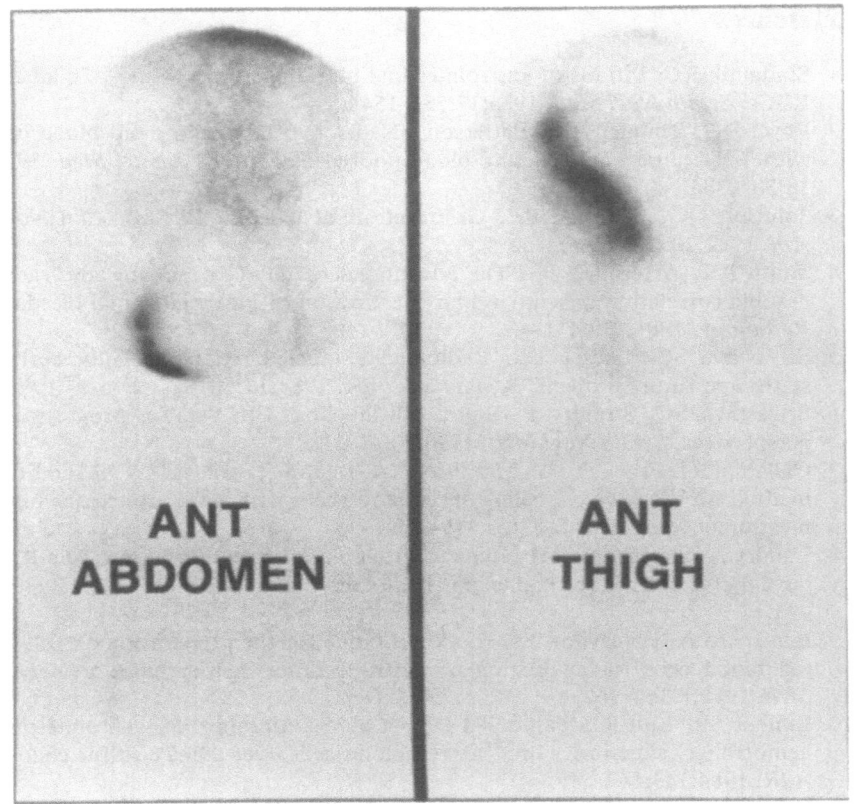

Figure 2.50. Bleeding after transfemoral arteriography.

An 83-year-old woman with a history of cirrhosis and chronic renal failure was admitted to the hospital for acute massive rectal bleeding. The patient underwent transfemoral arteriography with two injections of contrast into the abdominal aorta in order to locate the site of the GI bleeding. The test was negative. Two hours after transfemoral arteriography, $^{99m}TcO_4$ RBC abdominal imaging was performed (*left*). Scintigraphic detection showed an area of accumulation of the labeled RBCs in the upper third of the right thigh due to bleeding from the right femoral artery (*right*). The location corresponded to the site where the catheter was inserted for the angiography. Bleeding was not clinically obvious. No evidence for abdominal or esophageal bleed was found.[45]

(Reprinted from Gips S, Israel O. Scintigraphic detection of bleeding after transfemoral arteriography, using technetium 99m labeled RBCs. *Clin Nucl Med*. 1986; 11:669. With permission.)

References

1. Stadalnik RC. Pitfalls of gastrointestinal bleeding studies with ^{99m}Tc-labeled RBCs. *Semin Nucl Med.* 1986;17:151–154.
2. Pavel DG, Zimmer AM, Patterson VN. In vivo labeling of red blood cells with [99m]Tc: a new approach to blood pool visualization. *J Nucl Med.* 1977;18:305–308.
3. Johnson DG, Coleman RE. Gastrointestinal bleeding. *Radiol Clin North Am.* 1982;20:644–651.
4. Smith RK, Arterburn JG. The advantages of delayed imaging and radiographic correlation in scintigraphic localization of gastrointestinal bleeding. *Radiology.* 1981;139:471–462.
5. Srivastava SC, Chervu LR. Radionuclide-labeled red blood cells: current status and future projects. *Semin Nucl Med.* 1984;14:68–82.
6. Srivastava SC, Straub RF. Blood cell labeling with 99mTc: progress and perspectives. *Semin Nucl Med.* 1990;20:41–51.
7. Callahan RJ, Froelich JW, McKusick KA, Leppo J, Strauss HW. A modified method for the in vivo labeling of red blood cells with [99m]Tc: concise communication. *J Nucl Med.* 1982;23:315–318.
8. Landry A, Hartshorne MF, Bunker SR, et al. Optimal technetium-99m RBC labeling for gastrointestinal hemorrhage study. *Clin Nucl Med.* 1985;10:491–493.
9. Benedetto AR, Nusynowitz ML. A technique for the preparation of Tc-99m red blood cells for evaluation of gastrointestinal hemorrhage. *Clin Nucl Med.* 1983;8:160–162.
10. Bunker SR, Lull RJ, Tanasescu DE, et al. Scintigraphy of gastrointestinal hemorrhage: superiority of ^{99m}Tc red blood cells over 99mTc sulfur colloid. *AJR.* 1984;143:543–548.
11. Thorne DA, Datz FL, Remley K, Christian PE. Bleeding rates necessary for detection of acute gastrointestinal bleeding with technetium 99m-labeled red blood cells in an experimental model. *J Nucl Med.* 1987;28:514–520.
12. Smith R, Copley DJ, Bolen FH. 99mTc RBC scintigraphy: correlation of bleeding rates with scintigraphic findings. *AJR.* 1987;148:869–874.
13. Winzelberg GG, McKusick KA, Froelich JW, Callahan RJ, Strauss HW. Detection of gastrointestinal bleeding with ^{99m}Tc-labeled red blood cells. *Semin Nucl Med.* 1982;12:139–146.
14. Bunker SR, Hartshorne MF. Gastrointestinal hemorrhage. In: Mettler, FA ed. *Radionuclide Imaging of the GI Tract.* New York: Churchill Livingstone; 1986: 53–81.
15. Joseph UA, Jhingran SG. Technetium 99m-labeled RBC imaging in gastrointestinal bleeding from gastric leiomyoma. *Clin Nucl Med.* 1988;13:23–25.
16. Yen CK, Pollycove M, Parker H, Nalls G. Rupture of spontaneous aortoduodenal fistula visualized with Tc-RBC scintigraphy. *J Nucl Med.* 1983;24:332–333.
17. Tumeh SS, Parker JA, Royal HD, Uren RF, Kolodny GM. Detection of bleeding from angiodysplasia of the jejunum by blood pool scintigraphy. *Clin Nucl Med.* 1983;8:127–128.
18. Schneider M, Siegel ME. Gastrointestinal bleeding from an unusual etiology. *Clin Nucl Med.* 1984;9:243.
19. Winzelberg GG, Froelich JW, McKusick KA, et al. Radionuclide localization of lower gastrointestinal hemorrhage. *Radiology.* 1981;139:465–469.
20. Czerniak A, Zwas ST, Rabau MY, Avigad I, Borag B, Wolfstein I. Scintigraphic demonstration of acute gastrointestinal bleeding caused by gallbladder carcinoma eroding the colon. *Clin Nucl Med.* 1985;10:543–545.
21. Dorfman GS, Cronan JJ, Staudinger KM. Scintigraphic signs and pitfalls in lower gastrointestinal hemorrhage: the continued necessity of angiography. *RadioGraphics.* 1987;7:543–561.

22. Bunker SR, Brown JM, McAuley RJ, et al. Detection of gastrointestinal bleeding sites: use of in vitro technetium Tc 99m-labeled RBCs. *JAMA*. 1982;247:789–792.

23. Zuckier LS, Patel YD, Kratka PS, Sugarman LA. Pelvic kidney displacing a sigmoid diverticular bleed: scintigraphic and angiographic correlation. *Clin Nucl Med*. 1988;13:463–464.

24. Moreno AJ, Reeves TA, Pearson VD, Rodriguez Turnbull GL. Unusual manifestations of hemorrhage during technetium 99m red cell blood pool imaging. *Clin Nucl Med*. 1989;14:470–471.

25. Wise PA, Sagar VV. Retrograde flow pattern on a gastrointestinal bleeding scan. *Clin Nucl Med*. 1988;13:56.

26. Bakkers JTN, Crobach LFSJ, Pauwels EKJ. Factitious gastrointestinal bleeding. *J Nucl Med*. 1985;26:667–667.

27. Haseman MK. Potential pitfalls in the interpretation of erythrocyte scintigraphy for gastrointestinal hemorrhage. *Clin Nucl Med*. 1982;7:309–310.

28. Wilton GP, Wahl RL, Juni JE, Froelich JW. Detection of gastritis by ^{99m}Tc-labeled red-blood-cell scintigraphy. *AJR*. 1984;143:759–760.

29. Shih WJ, Magoun S, Donstad PA. Huge abdominal aortic aneurysm demonstrated by technetium 99m labeled RBC blood pool imaging. *Clin Nucl Med*. 1987;12:825–826.

30. Fritsch S, Arleth J, Becker HM, Mannes AG, Moser E. Detection of a false aneurysm in the thigh by scintiangiography: case report. *Clin Nucl Med*. 1987;12:564–566.

31. Shih WJ, Domstad PA, Loh FG, Pulmano C. Extensive gastric varices demonstrated by technetium 99m red blood cell scintigraphy. *Clin Nucl Med*. 1987;12:290–293.

32. Iles S, Gordon D, Aquino JA, Caines J. Visualization of portal venous collaterals on a technetium 99m-labeled red blood cell scan for gastrointestinal bleeding: potential pitfall in interpretation. *Clin Nucl Med*. 1987;9:946–948.

33. Camele RA, Bansal SK, Turbiner EH. Red blood cell gastrointestinal bleeding scintigraphy: appearance of the left ovarian vein. *Clin Nucl Med*. 1984;9:275–276.

34. Moreno AJ, Byrd BF, Berger DE, Turnbull GL. Abdominal varices mimicking an acute gastrointestinal hemorrhage during technetium 99m red blood cell scintigraphy. *Clin Nucl Med*. 1985;10:248–251.

35. Howard JL, Dhekne RD, Moore WH, Long SE. Double inferior vena cava by technetium 99m labeled RBC study. *Clin Nucl Med*. 1988;13:671–672.

36. Holoway H, Johnson J, Sandler M. Detection of an ileal cavernous hemangioma by technetium 99m red blood cell imaging. *Clin Nucl Med*. 1988;13:32–35.

37. McDonald K. Technetium 99m RBC scintigraphy in the evaluation of small bowel leiomyoma. *Clin Nucl Med*. 1987;12:131–133.

38. Front D, Ora I, Groshar, Weininger J. Technetium 99m-labeled red blood cell imaging. *Semin Nucl Med*. 1984;14:226–250.

39. Wood MJ, Hennigan DB. Radionuclide tagged red blood cells in the gallbladder. *Clin Nucl Med*. 1984;9:289–290.

40. Lee SM, Lee RG, Clouse ME, Hill TC. Demonstration of hematobilia using technetium 99m labeled red blood cells. *Clin Nucl Med*. 1986;11:52.

41. Ellison MJ, Thornburg A, Turbiner E. Demonstration of bleeding into a pancreatic pseudocyst on a technetium 99m labeled red blood cell scan. *Clin Nucl Med*. 1987;12:969.

42. Mitchell DS, Stacy TM, Grunow JE, Leonard JC. Scintigraphic detection of an occult abdominal bleed in a child. *Clin Nucl Med*. 1988;13:546–548.

43. Czarnecki DJ. Intraperitoneal hemorrhage diagnosed by technetium 99m labeled RBC imaging. *Clin Nucl Med*. 1986;11:617–618.

44. Fink-Bennett D, Johnson JR. Gluteal hematoma: a potential cause for a

false-positive Tc-99m RBC gastrointestinal bleeding study. *Clin Nucl Med.* 1984;9:414.

45. Gips S, Israel O. Scintigraphic detection of bleeding after transfemoral arteriography, using technetium 99m labeled RBCs. *Clin Nucl Med.* 1986;11: 669.

CHAPTER 3

Atlas of ^{99m}Tc Labeled Red Blood Cell Liver Scintigraphy

Harvey A. Ziessman

Technetium-99m labeled red blood cell (^{99m}Tc RBC) scintigraphy has been found extremely useful for the diagnosis of cavernous hemangiomas of the liver. This atlas chapter will briefly review the pathophysiology, clinical manifestations, and importance of these benign tumors. The diagnostic role of other imaging modalities, such as ultrasonography, CT, and MRI will be discussed. Emphasis will be placed on the overall accuracy and clinical utility of ^{99m}Tc RBC planar and SPECT scintigraphy. Examples in the Atlas Section will emphasize important diagnostic teaching points.

Cavernous hemangiomas of the liver are the the most common benign tumor of the liver and the second most common hepatic tumor, exceeded in incidence only by hepatic metastases.[1] Histologically, these tumors comprise endothelial-lined vascular channels of varying sizes, separated by fibrous septa. They are usually small and single, but may be large and occupy most of the liver. They are not related pathologically to capillary hemangiomas, infantile hemangioendotheliomas, or angiodysplastic lesions. Ten percent of hemangiomas are multiple. Lesions larger than 4 cm are frequently referred to as giant cavernous hemangiomas. The autopsy incidence has been reported to be between 0.4% and 7.0%. They are more common in women by a ratio of 4–6 : 1.[2–4]

Symptoms from hemangioma of the liver are uncommon. Only rarely do patients present with hepatomegaly or abdominal discomfort due to pressure symptoms on adjacent organs. More acute pain can be due to thrombosis, infarction, or very rarely rupture and hemorrhage.[1,3] Liver function tests are normal unless elevated by concomitant disease.

The natural history of these benign tumors is not well defined since they are usually asymptomatic and are discovered incidentally. However, they may increase in size with pregnancy or age and undergo varying degrees of degeneration, fibrosis, and calcification.

Hemangiomas are frequently detected incidentally by sonography or computed tomography (CT) during the clinical workup or staging of a patient with a known primary malignancy, or during evaluation of unrelated abdominal symptoms or disease.

The classic sonographic pattern of a homogeneous, hyperechoic mass with well defined margins and posterior acoustical enhancement is neither sensitive nor specific for the diagnosis of hemangioma.[5–7] Strict criteria for the classic CT appearance include relative hypoattenuation before intravenous contrast injection, early peripheral enhancement during the rapid bolus dynamic phase of contrast administration, progressive opacification toward the center of the lesion, and complete isodense fill-in usually by 30 min after contrast administration. Frequently all criteria are not satisfied. When these criteria are used to maximize specificity, the sensitivity of CT is only 55% to 76%.[8,9] Therefore, up to 45% of hemangiomas may not be diagnosed using these strict CT criteria. Less strict criteria result in increasing false positives. Recent preliminary reports have claimed a high accuracy for the diagnosis of hepatic hemangiomas with magnetic resonance imaging (MRI). Hemangiomas have a characteristic appearance on MRI with high signal intensity on T2-weighted spin-echo images. Although MRI is more accurate than CT or sonography, a variety of other benign and malignant tumors have been reported to give false positive results, including metastatic adenocarcinoma of the lung, metastatic carcinoid, pheochromocytoma, islet cell carcinoma, pancreatic and uterine adenocarcinoma, and various sarcomas.[5,10-13] As would be expected, cavernous hemangiomas have nonspecific decreased uptake on ^{99m}Tc sulfur colloid (^{99m}Tc SC) liver–spleen scans. Angiography is usually diagnostic, but invasive and usually unnecessary.[4]

^{99m}Tc RBC liver scintigraphy has been found to be an extremely accurate method for the diagnosis of cavernous hemangiomas of the liver.[14-16] It is a highly specific test and only a few false positive studies have been reported. Four hepatomas[14,17,18] and one hepatic angiosarcoma[19] have been reported to have increased uptake on 1- to 2-hr delayed imaging, the diagnostic pattern of hemangiomas. However, hepatic angiosarcomas are extremely rare. Although some authors have emphasized the potential for false positive studies due to hepatomas, in fact most hepatomas have negative ^{99m}Tc RBC liver scintigraphy. In one recent series that included 46 hepatomas, none had the pattern of increased uptake on delayed imaging seen with hemangiomas,[16] so the specificity of ^{99m}Tc RBC scintigraphy is very high, approaching 100%.

The sensitivity of ^{99m}Tc labeled RBC liver scintigraphy depends on lesion size and the camera technology used. Although extensive fibrosis or thrombosis may result in false negative studies, this is uncommon. Planar studies can typically detect hemangiomas 3 cm and larger in size whereas single photon emission computed tomography (SPECT) is able to detect hemangiomas down to a size of 2 cm or less.[20-22] SPECT is particularly useful for identifying centrally located lesions and lesions adjacent to the heart, major vessels, spleen, or kidney. This is an area where SPECT imaging has a clear superiority over planar imaging. With improved SPECT technology, hemangiomas as small as 1 cm can now be detected.[21,22] Kudo et al. found a sensitivity of 43% for planar imaging and 83% for SPECT in lesions sized 2.1 to 3.0 cm using a single-headed rotating gamma camera.[15] Although they could detect lesions as small as 1 cm, SPECT sensitivity was only 38% for hemangiomas ≤2 cm in size. At Georgetown University Hospital, using a high resolution three-headed rotating SPECT camera (Triad, Trionix Research Laboratory, Inc.), we have found an improved ability to detect small and multiple hemangiomas compared to our single-headed tomographic cameras.[23] In a recent review of 18 consecutive patients (38 lesions) referred for evaluation of uncertain lesions seen on ultrasound, CT, or MRI, we have found a high sensitivity (100% to date) for diagnosis of lesions 1.4 cm and larger and have visualized hemangiomas down to a size of 5 mm, although our sensitivity for these smaller lesions is less.[24]

Few direct comparisons of ^{99m}Tc RBC SPECT and MRI have been performed.[13,25] The present data suggest that ^{99m}Tc RBC liver scintigraphy is more specific (has fewer false positives), but that MRI is more sensitive for detecting lesions smaller than 2 cm in size, and particularly small lesions adjacent to a major intrahepatic vessel.[13]

^{99m}Tc RBC scintigraphy can usually obviate the need for invasive angiography or biopsy. Unknowing biopsy of a cavernous hemangioma has resulted in serious hemorrhage and even death.[26,27] Since these benign tumors do not cause symptoms or morbidity, no specific therapy is indicated.

Hemangiomas of the liver in infants and children (capillary hemangiomas and infantile hemangioendotheliomas) present differently from cavernous hemangiomas in adults. They are rarely asymptomatic and typically present with hepatomegaly and congestive heart failure, frequently simulating a primary hepatic malignancy clinically. They may simulate malignancy on cross-sectional imaging or angiography as well. ^{99m}Tc RBC liver scintigraphy is diagnostic.[28]

Technique

The following protocol will describe a combined three-phase planar and SPECT technique that we have used for ^{99m}Tc RBC scintigraphy. However, it is our feeling that SPECT alone is diagnostic of liver hemangiomas.[24]

Imaging Procedure

Patient preparation: The patient's CT scan, ultrasound, MRI, or ^{99m}Tc SC liver–spleen study should be used to determine in which projection the flow study should be performed and for correlative image interpretation. The chosen flow view should give optimal visibility to the lesion. We perform a ^{99m}Tc SC SPECT study on a separate day only if CT or MRI is not available. The correlative study, whether CT, MRI, or SPECT ^{99m}Tc SC study scan, is vital for anatomical localization and accurate ^{99m}Tc RBC correlation and interpretation.

Radiopharmaceutical: ^{99m}Tc pertechnetate, 25 mCi. We use the modified in vivo technique of red blood cell labeling.[29]

Instrumentation: Camera: Large field of view gamma camera with tomographic capability. Collimator: High resolution low energy parallel hole. Energy window: 15% centered over a 140-keV photopeak.

Aquisition protocol: Planar three-phase imaging.
1. Blood flow: Acquire 1-sec frames for 60 sec on computer and 2-sec frames on analog film.
2. Inject the ^{99m}Tc labeled red blood cells as an intravenous bolus.
3. Immediate images: Immediately after the flow study, acquire a 1 million count planar image in the same projection. Acquire other views as deemed necessary to visualize best the lesion in question. Oblique images are frequently helpful.
4. Delayed images: Acquire planar static images 1 to 2 hr after injection in the same manner and order as described for immediate images.
5. SPECT: The exact technique will depend on your particular instrumentation, computer, software, and personal preferences. The following is our protocal for using a single-headed rotating tomographic

gamma camera (Siemen's 7500 and Siemen's Microdelta computer and software) as well as our protocol for our three-headed dedicated SPECT system (Triad, Trionix).

SPECT

Patient Set-Up

1. Position the patient supine on the imaging table. Raise the patient's arms over his head. Place a pillow under the knees to reduce low back discomfort.
2. Center the liver in the field of view.
3. Rotate the camera head around the patient to ensure that the camera does not come in contact with the patient. Observe the organ of interest on the camera or computer display. It should remain in the field of view during a test rotation.

Single-Headed Rotating SPECT

Camera Set-Up

Window: 15% window centered over 140-keV ^{99m}Tc photopeak
Set-up: Step and shoot
Collimator: High resolution

Computer Set-Up

Acquisition parameters:
Patient orientation: supine
Rotation: clockwise
Matrix: 64 × 64 word mode
Image/Arc combination: 128 images/360°
Time/Frame: 10 sec/frame
Reconstruction parameters:
Interslice filter: 1,2,1 smoothing filter
Convolution filter: Shepp–Logan Hamming; Cutoff 0.8 Nyquist frequency; Alpha weighting, 0.5
Attenuation correction: Chang
Reformatting: sagittal/coronal

Triple-Headed Dedicated SPECT System

Camera Set-Up

Window: 15% window centered over 140-keV ^{99m}Tc photopeak
Set-up: Step and shoot, noncircular orbit
Collimators: Ultrahigh resolution

Computer Set-Up

Acquisition parameters:
 Patient orientation: supine
 Rotation: clockwise as viewed from the patient's feet
 Matrix: 128 × 128 word mode
 Image/Arc combination: 40 images/120° for each detector. Total of
  120 images/360°
 Time/Frame: 40 sec/image
Reconstruction parameters:
 Interslice filter: 1,3,5,3,1
 Convolution filter: Hamming, 1.0 cycles/cm
 Attenuation correction: Chang 0.11 attenuation coefficient
 Reformatting: sagittal/coronal

Physiological Basis

Although hepatic artery branches feeding the hemangioma are frequently displaced and crowded to the side by the hemangioma, which
behaves like a benign space-occupying mass, the vessels tend to be small,
are not enlarged, taper normally, and divide into normal small vessels
before filling the enlarged vascular spaces.[4] As a result, the arterial phase
on both contrast and radionuclide angiography is usually normal. Cavernous hemangiomas of the liver are in reality large blood-filled cavities,
essentially large varix-like spaces. With time, the labeled red blood cells
equilibrate within these abnormally dilated vascular channels. This process of equilibration within the liver hemangioma is slower than other
areas of increased uptake on ^{99m}Tc RBC imaging; for example, the heart
chambers, which empty and replenish a large portion of their blood pool
with each cardiac cycle, and the spleen and kidney, which are richly
vascularized organs with normal capillaries and venules and, as a result,
have a rapidly changing blood pool. The characteristic finding on ^{99m}Tc
RBC scintigraphy of initially decreased activity compared to adjacent
liver and delayed increased uptake is explained by the fact that, although
the normal liver is vascularized quickly with labeled red blood cells via
the hepatic artery and portal vein, it takes time for the newly labeled
cells to equilibrate within the large, relatively stagnant blood pool in
the hemangioma. The percent of labeled cells, initially small, gradually
increases as new labeled blood replaces unlabeled blood. Finally, when
equilibrium of the labeled cells within the blood pool is achieved, the
relative uptake within the hemangioma will be equal to heart blood pool
and greater than the less well vascularized adjacent normal liver.

Visual Description and Interpretation

The diagnostic scintigraphic pattern of cavernous hemangioma is that of
increased uptake compared to adjacent liver on delayed imaging 1 to 2
hours after injection. Except for the rare exceptions described in the
introduction, all other benign and malignant liver tumors (e.g., metasta-

Table 3.1. Estimated radiation absorbed dose.

Target	rad/mCi	rads/25 mCi	mGy/MBq
Heart wall	0.057	1.4	0.015
Bladder wall	0.087	2.2	0.024
Spleen	0.043	1.1	0.012
Blood	0.038	1.0	0.010
Liver	0.028	0.7	0.008
Kidneys	0.027	0.7	0.007
Red marrow	0.020	0.5	0.005
Thyroid	0.019	0.5	0.005
Ovaries	0.018	0.5	0.005
Testes	0.008	0.2	0.002
Total body	0.016	0.4	0.004

Adapted with permission from Atkins et al., ref. 30.
The calculations are based on a 4.8-hr voiding schedule, and using the calculated radiation absorbed dose that was highest from either in vivo or in vitro labeling.

ses, abcesses, cysts, and cirrhotic nodules) have decreased activity.[14] Occasional hemangiomas will show heterogeneity of uptake with areas of both increased and decreased uptake and some may have a central cold area. These patterns are usually attributed to partial thrombosis or fibrosis.[31,32] There have been a few reports of uptake in hemangiomas equal to liver background.[33,34] However, the authors believe that this pattern is not diagnostic for hemangioma, since a similar pattern was seen with lesions other than hemangioma (e.g., metastasis and hepatoma). These reports have been rare and only planar imaging was used.

Three-phase studies has been widely recommended with initial blood flow and immediate static imaging, followed by the delayed 1- to 2 hour imaging.[14,28,33] The blood flow phase is typically normal, although increased and decreased flow has been described.[14,33] Although some previous reports have stated that increased flow suggests hepatocellular carcinoma, Kudo et al., studying a large number of patients with hepatomas and hemangiomas <5 cm in size, found a similar incidence of increased flow in both entities. In fact, most hemangiomas and hepatomas had normal flow. That study did not find the flow phase of the study helpful in making the differential diagnosis.[16] The postflow immediate images characteristically show a cold defect that fills in on delayed imaging. However, this pattern is variable and early filling-in is frequently seen.[32,33] The added value of performing routine flow and immediate images is questionable. Delayed SPECT imaging is almost always diagnostic by itself.

SPECT has shown clear superiority over planar imaging for the diagnosis of cavernous hemangomas.[16,20,35,36] The improved image contrast, and therefore lesion detectability, make it the technique of choice. Delayed SPECT imaging alone is probably all that is needed for accurate diagnosis. However, it is important to review the rotating cinematic display and all cross-sectional slices (transverse, coronal, and sagittal) and correlate the results carefully with the patient's sonography, CT, MRI, or SPECT ^{99m}Tc SC study. Careful delineation of normal vascular structures by sequencing through consecutive slices will prevent misdiagnosis.[37,38] The atlas section will emphasize important diagnostic and teaching points.

Miller has found ^{99m}Tc RBC scintigraphy to be diagnostic in differentiating infantile hemangioendotheliomas and childhood capillary hemangiomas from maliganant disease.[28] Different from the usual adult liver hemangomas, these benign childhood vascular tumors almost invariably have increased flow, increased activity on immediate imaging, and uptake within the lesions equal to heart blood pool on delayed imaging. Although benign and malignant lesions other than hemangiomas may have activity equal to liver, Miller did not find any patients with uptake equal to heart who did not have a hemangioma. No false positives or false negatives were found.

Acknowledgments. The author wishes to thank Beth Harkness and Valerie Swineford for their technical assistence and John Patterson, M.D., John W. Keyes, Jr., M.D., and Paul Silverman, M.D., for their valuable help and advice in making this atlas possible.

Atlas Section

Figure 3.1. Classical findings: giant cavernous hemangioma.

A: Planar ^{99m}Tc RBC imaging 1 hr after labeling shows a large hemangioma involving the inferior aspect of the liver.
Comment: Uptake within this large tumor is greater than liver and equal to heart and spleen. Note the relatively inhomogeneous uptake with decreased uptake inferolaterally, probably due to areas of degeneration and fibrosis.
B: Contrast-enhanced CT slice through the center of this giant cavernous hemangioma for comparison also shows the inhomogeneity of this benign vascular tumor.
Comment: The CT scan suggested hemangioma; however, strict criteria for hemangioma were not met.

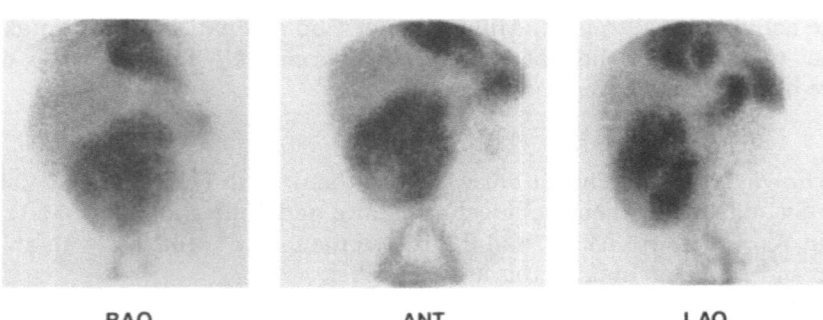

RAO   ANT   LAO

A

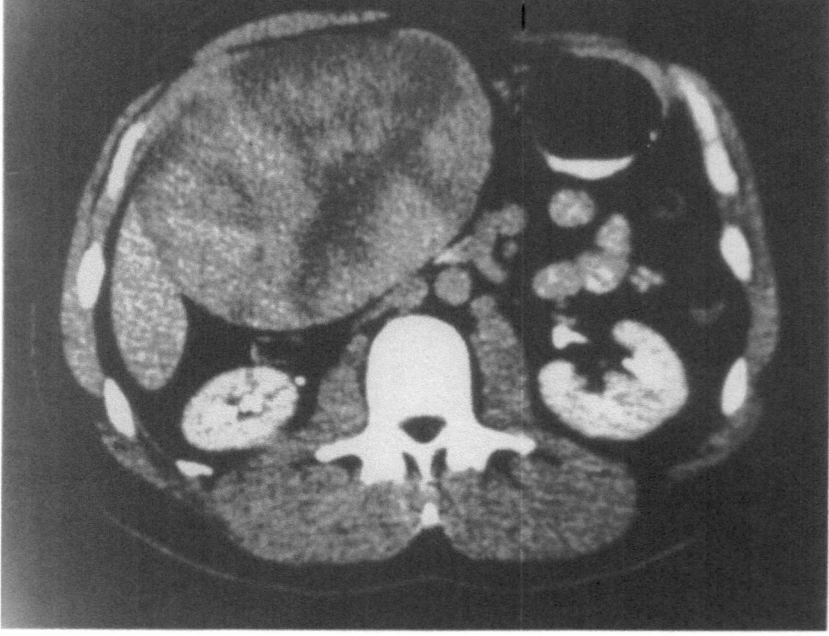

B

IMMED   DELAYED

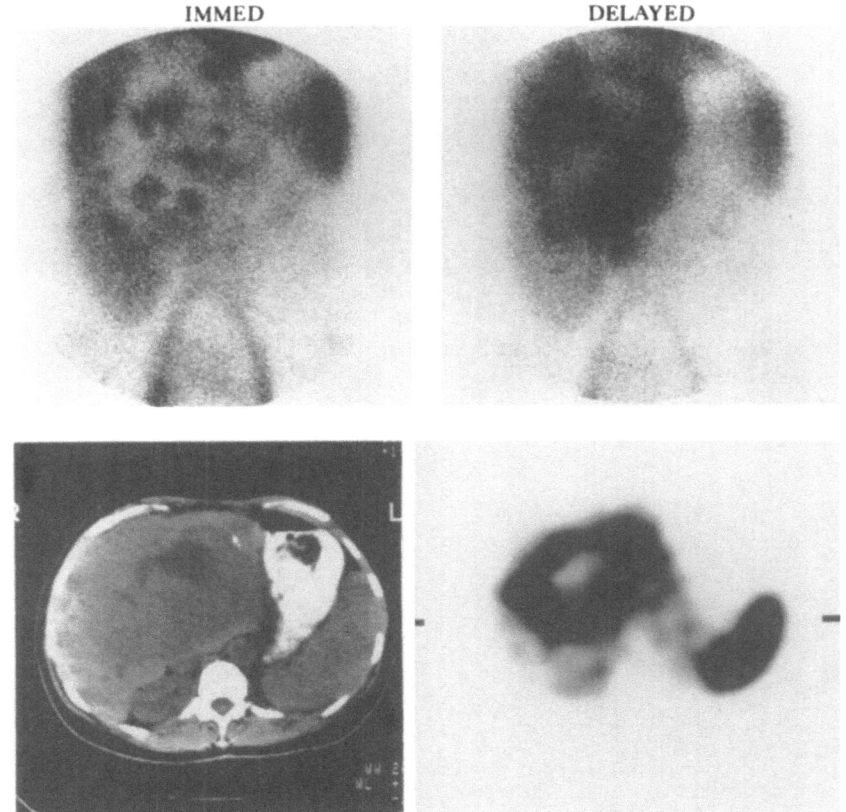

A

B

Figure 3.2. Classical findings: large cavernous hemangioma.

A: Planar. An immediate anterior image (*left*) shows a large cold defect involving the left lobe and a large portion of the right lobe. Small areas of early filling-in can be seen. The 2-hr delayed image (*right*) shows continued and near complete filling-in of this large hemangioma. A few central areas of decreased uptake are seen.

B: CT (*left*): A selected transaxial CT slice shows the patient's large lesion. Although suspicious for a hemangioma, it did not have all the classical findings. SPECT (*right*): This is a transverse 6-mm cut at approximately the same level as the CT. The uptake is intense and equal to the adjacent spleen. The central cold area on SPECT corresponds to a similar area on the CT.

Comment: The central cold area is due to either fibrosis, thrombosis, or hemorrhage. Very large hemangiomas may sometimes be more obvious on planar than SPECT imaging since cross-sectional slices may include only hemangioma and no normal liver for comparison.

Figure 3.3. Classical findings for hemangioma: three-phase planar and SPECT ^{99m}Tc RBC study.

A: Flow study (2 sec/frame) shows normal arterial (A) phase flow to the lesion in the lateral aspect of the right lobe of the liver. Increased venous (V) phase flow is noted at the periphery of the lesion, particularly medially, and a cold center.

B: Planar. The immediate anterior image (*left*) shows a cold defect in the lateral aspect of the right lobe. However, early filling-in can be seen superiorly (*arrowhead*). The 1-hr delayed image (*right*) shows complete filling-in, with activity greater than liver and equal to heart, diagnostic of a hemangioma.

Comment: Figures 3.3A–B illustrate the classical planar findings for hemangioma. Normal flow is seen most commonly, although increased or decreased flow may be seen. Although the immediate images usually show a cold defect, early filling-in is common. The delayed planar images show the typical findings of increased uptake, which is greater than adjacent liver and equal to heart.

C,D,E: SPECT. Selected transverse (C), coronal (D), and sagittal (E) cross-sectional slices performed with a single-headed rotating gamma camera. For all three figures, the above two images are consecutive 6-mm ^{99m}Tc SC liver–spleen slices and below are comparable cuts of the ^{99m}Tc RBC study.

Comment: SPECT gives excellent contrast resolution; however, it probably does not add much diagnostically for interpretation of this large (8 cm) hemangioma.

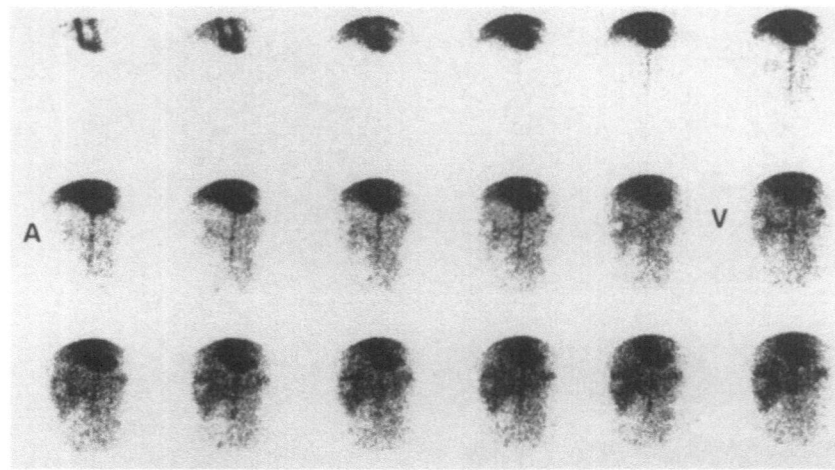

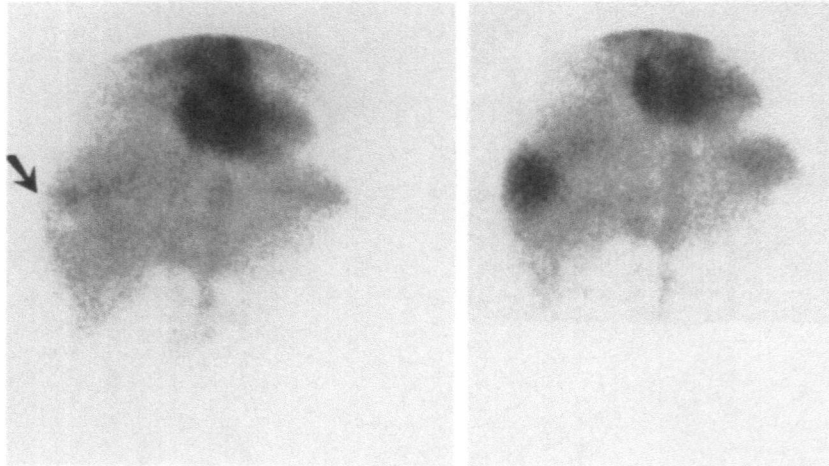

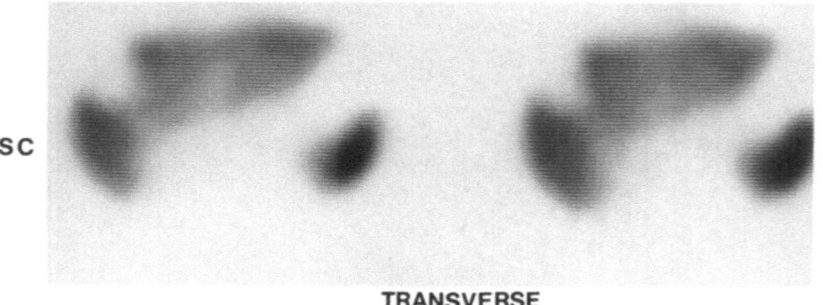

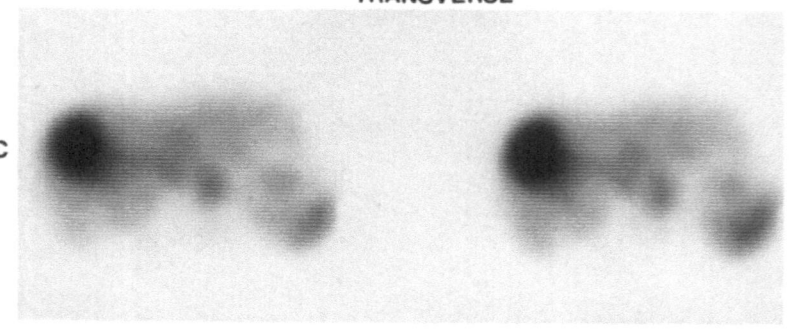

TRANSVERSE

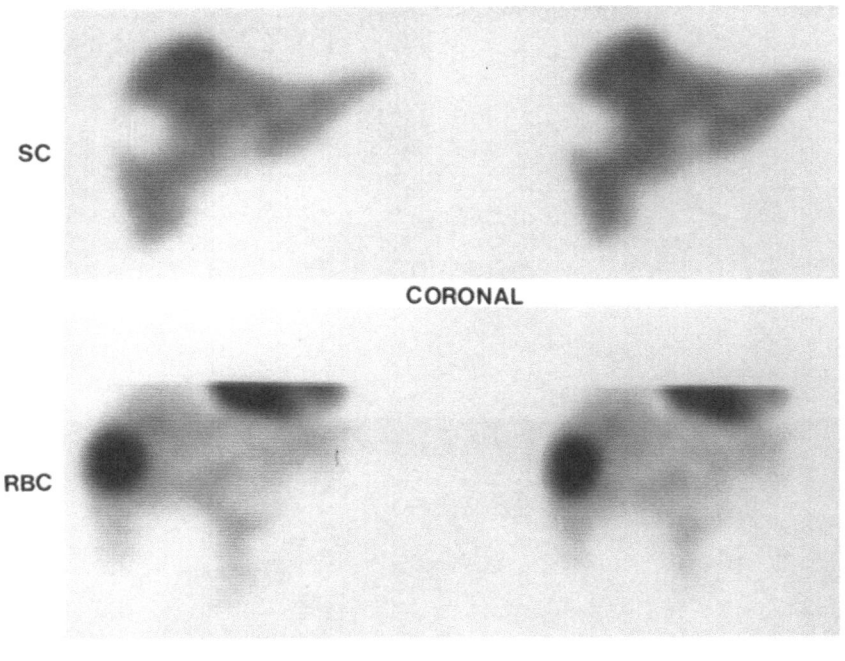

SC

CORONAL

RBC

D

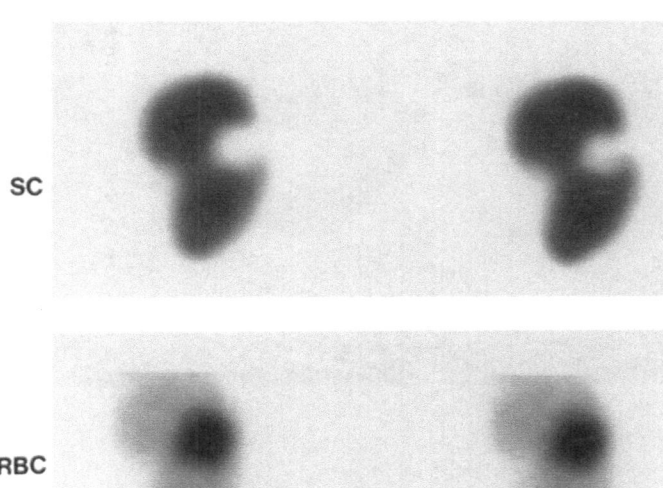

SC

RBC

E

Figure 3.4. Value of ^{99m}Tc RBC scintigraphy: hemangioma-negative study.

The CT scan (*right*) shows a large lesion in the midliver that was interpreted as likely being a benign cyst or possibly a hemangioma. The ^{99m}Tc RBC 1-hr delayed SPECT study (*left*) is cold in the region of the CT lesion and, therefore, negative for hemangioma. The ^{99m}TC RBC study excluded the diagnosis of hemangioma.

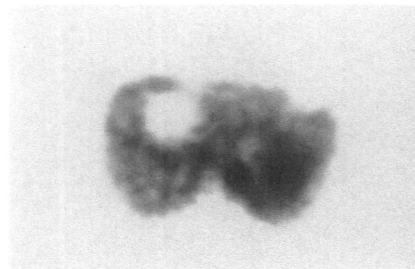
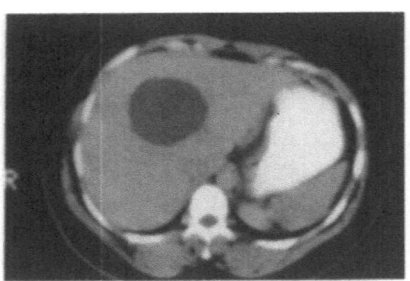

Figure 3.5. Value of ^{99m}Tc RBC scintigraphy: hemangioma-negative study in patient with metastatic colon cancer.

A: CT study shows a lesion in the posterior aspect of the right lobe suspicious for metastatic cancer (*arrow*). The patient was referred to rule out a hemangioma before biopsy.
B: Two sequential 6-mm transaxial SPECT cuts at approximately the same level as the CT scan in **A** show a cold area (*arrow*) that corresponds to the lesion seen on CT scan. This is negative for hemangioma. Subsequent biopsy of this lesion was diagnostic of adenocarcinoma metastatic to the liver.
Comment: Inadvertent biopsy of a hemangioma can result in serious hemorrhage.

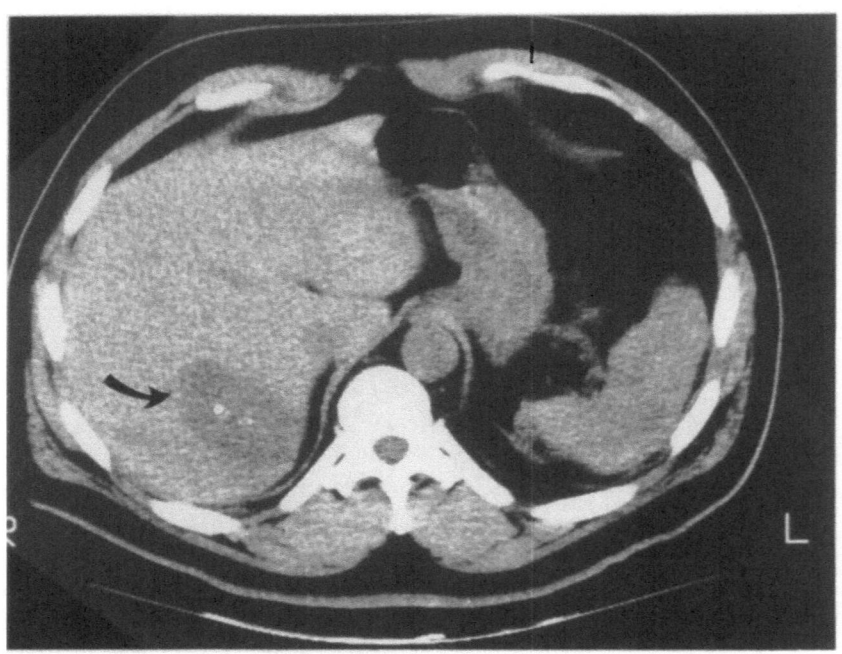

A

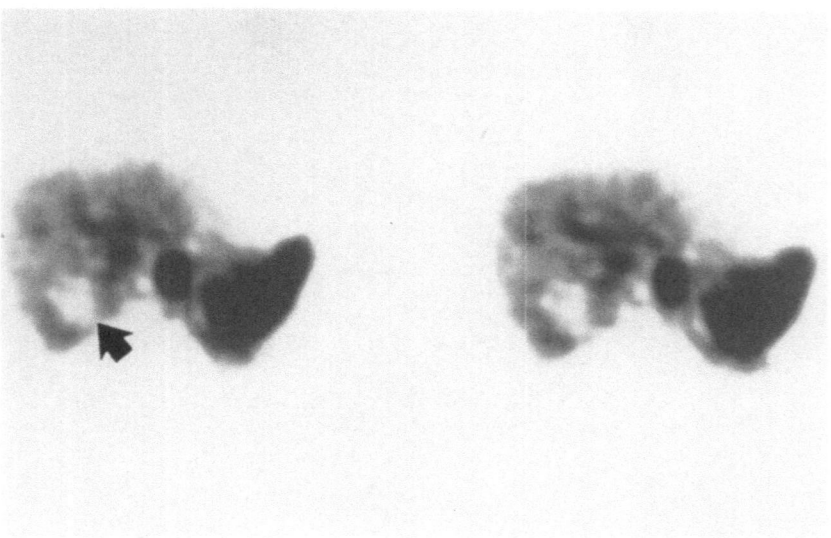

B

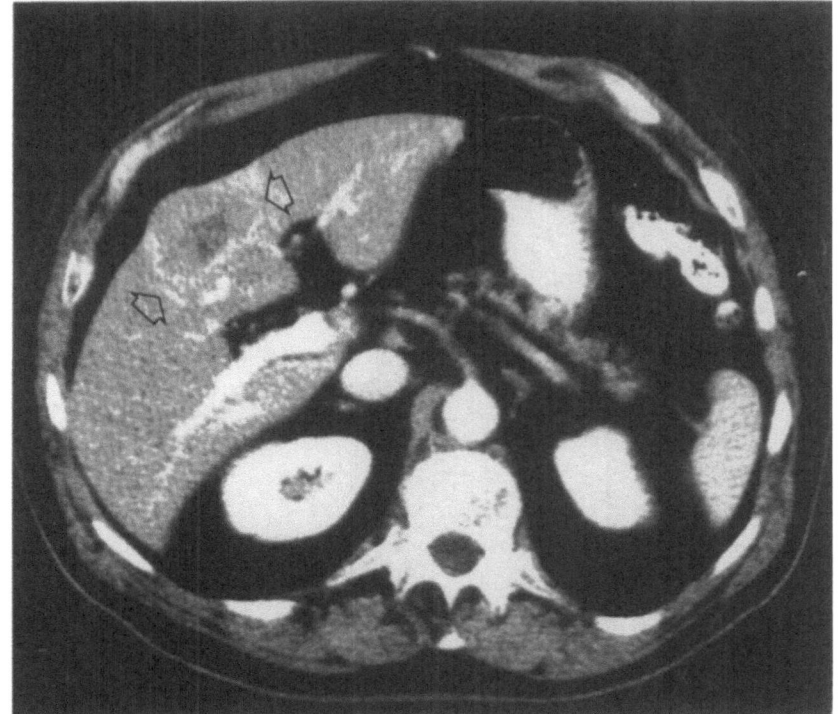

Figure 3.6. Value of ^{99m}Tc RBC scintigraphy: hemangioma-negative study in a patient with metastic cancer.

A: A contrast-enhanced CT scan shows a lesion (*arrowheads*) interpreted as probable metastatic cancer. The patient had a history of colon cancer and was referred for ^{99m}Tc RBC scintigraphy to exclude a hemangioma.

B: Two sequential transaxial slices cut at the same level as the CT scan in **A**. The lesion is negative for hemangioma (*open arrowheads*). Note the prominent portal vein (P) due to portal hypertension. The kidney (K), inferior vena cava (I) and aorta (A) are noted. Biopsy confirmed metastatic colon cancer to the liver.

Comment: The hemangioma-negative lesions are less distinct here than seen on the examples in Figures 3.4 and 3.5. Careful comparison of the sequential CT and SPECT slices is often necessary to ensure proper interpretation.

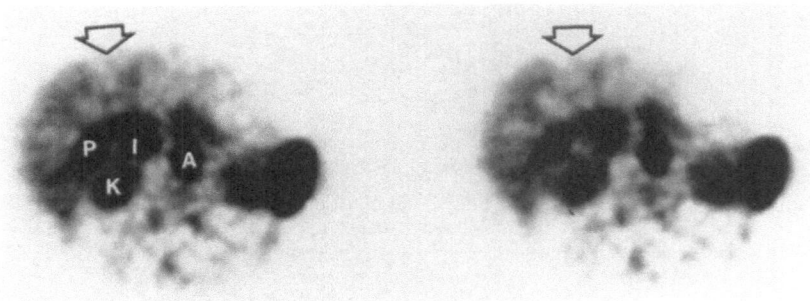

A

B

Figure 3.7. Value of ^{99m}Tc RBC scintigraphy: hemangioma-negative and hemangioma-positive lesions in the same patient study.

A: Two selected CT slices show a contrast-enhanced lesion in the inferior aspect of the right lobe (*left*) suspicious for a hemangioma and a second lesion in the superior aspect of the left lobe (*right*) consistent with a cyst.

B: The planar study acquired at 1 hr shows a hemangioma in the anterolateral aspect of the right lobe (*arrowhead*) consistent with one of the lesions seen on CT. However, the second cystic-appearing lesion on CT cannot be defined on planar imaging.

C: A selected coronal SPECT slice is shown. Both lesions can be seen: the hemangioma as increased uptake inferolaterally in the right lobe (*arrowhead*) and the cyst as a cold defect immediately adjacent to the heart (*white arrow*) and, therefore, hemangioma-negative.

Comment: This study, which was acquired with a single-headed tomographic gamma camera, demonstrates the improved contrast resolution of SPECT. The planar study (**B**) was not able to resolve the cold lesion immediately adjacent to the heart.

D: The three-view cross-sectional SPECT slices, transverse (TRANS), coronal (COR), and sagittal (SAG), clearly show the cold lesion to be anterior (A) and adjacent to the heart (H); (P = posterior).

Comment: Multiple projections are helpful in localizing a lesion in three dimensions. Cold and hot lesions may be quite difficult to define on planar imaging. The ability of SPECT to remove overlying and underlying activity improves the contrast resolution.

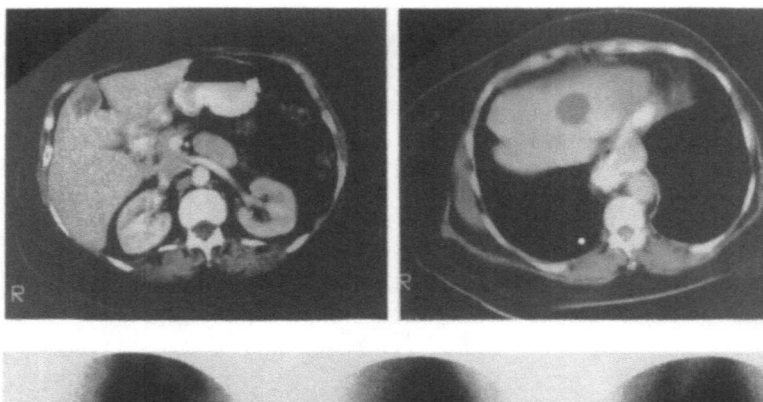

A

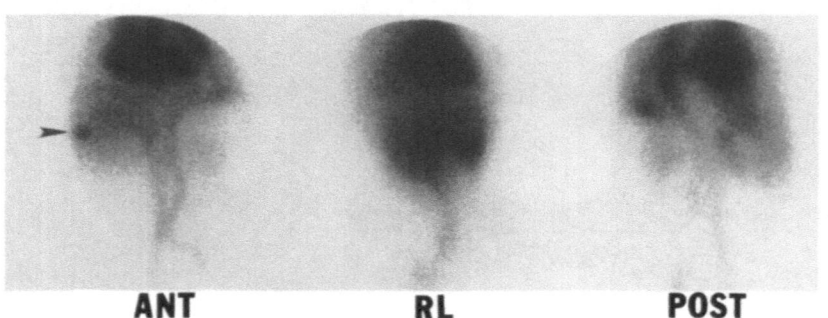

ANT   **RL**   **POST**   B

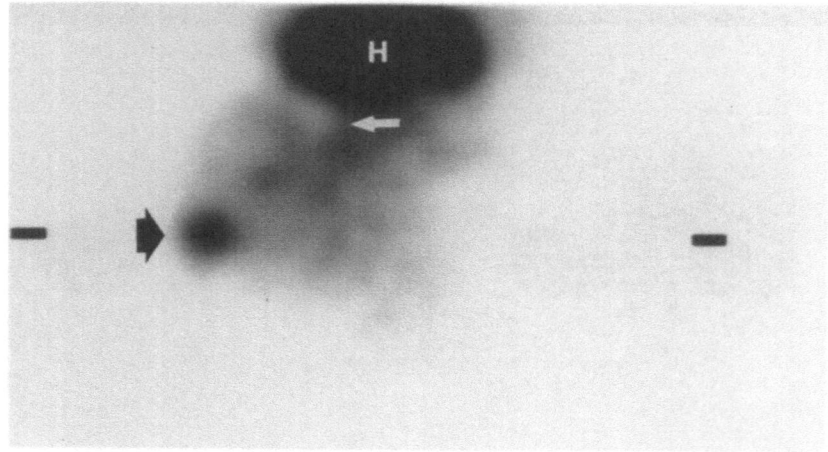

C

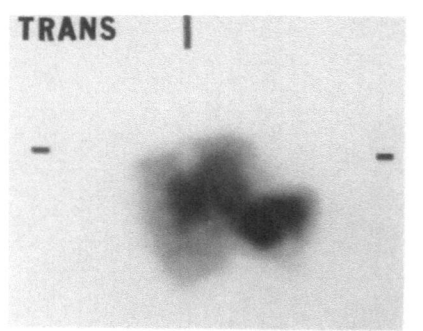
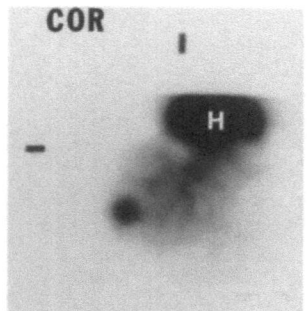
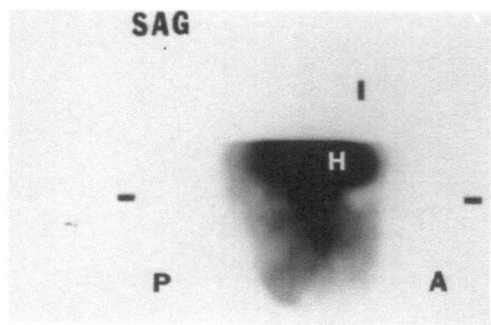

D

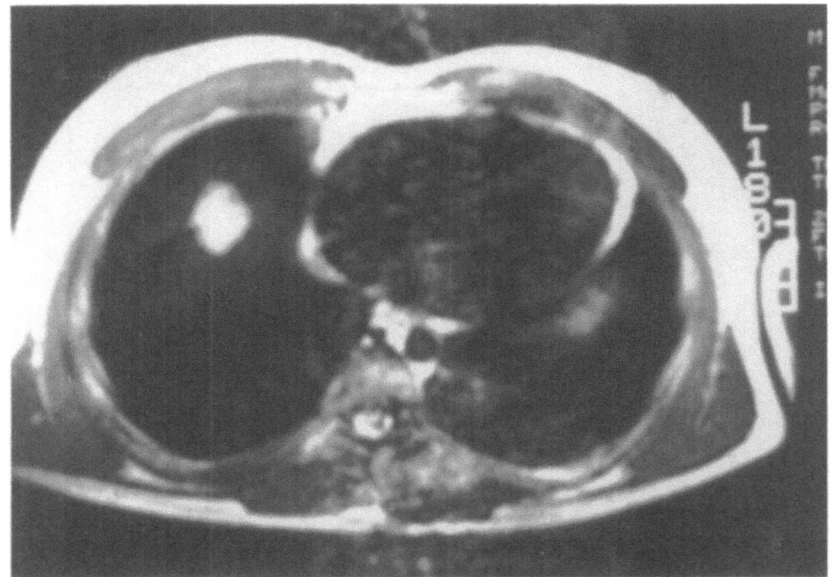

A

B

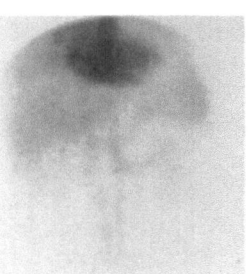

IMMEDIATE

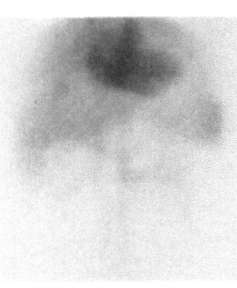

30 MIN

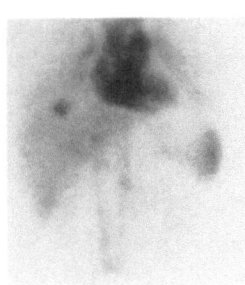
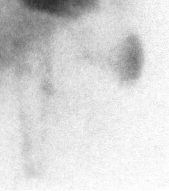

2 HR

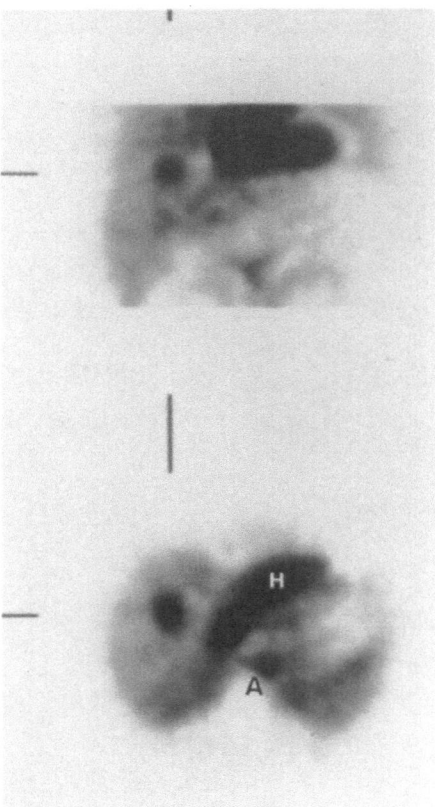

C

Figure 3.8. Importance of adequate delay for planar imaging: advantage of SPECT.

A: T2-weighted MRI image shows a lesion in the dome of the right lobe of the liver with high signal intensity suggestive of a hemangioma.

Comment: Since false positives due to a number of benign and malignant tumors occur with MRI, the patient was referred for the more specific test, ^{99m}Tc RBC scintigraphy.

B: From top to bottom are an immediate, 30-min, and 2-hr postinjection anterior images. It is difficult to define the lesion on the immediate image. The 30-min image shows only minimal and subtle increase in activity greater than background in the dome of the right lobe, suggestive but not diagnostic of hemangioma. Delayed imaging at 2 hr is definitive.

Comment: Imaging too early may miss the increased uptake of a hemangioma. Adequate delay may be particularly important with small and centrally located lesions.

C: SPECT study (coronal slice above and transverse below) acquired about 1 hr after injection (after the 30-min, but before the 2-hr planar study in **B**) is clearly positive for hemangioma in the dome of the liver.

Comment: Note the improved contrast with SPECT with a very high lesion to liver activity ratio compared to planar imaging. For this reason, SPECT imaging may be diagnostic earlier (e.g., 30 min) compared to planar imaging (1–2 hr). However, SPECT imaging at 1 hr is recommended, especially for smaller lesions. The heart (H) and aorta (A) are noted.

Figure 3.9. Value of SPECT over planar imaging: normal vascular anatomy.

A: Three-view planar study shows a hemangioma in the anterolateral aspect of the right lobe, seen in the anterior view, but having the highest hemangioma-to-liver-uptake ratio in the posterior view (*arrowhead*). A right posterior oblique view may have shown it best. It is not seen well in the lateral view, proably due to adjacent/overlying vascular structures.

B: Two sequential transverse, coronal, and sagittal views cutting through the hemangioma. Note the much improved contrast resolution compared to the planar study.

C,D: Sequential 6-mm coronal sections from anterior (**C,D**) to posterior (**C,D**).

Comment: Sequencing through consecutive slices is helpful in differentiating normal from abnormal vascular structures. The vascular anatomy can sometimes be followed more easily in the coronal slices. The following vascular structures are labeled: portal vein (PV), aorta (Ao), inferior vena cava (IVC), hepatic vein (HV), hemangioma (H), and kidney (K).

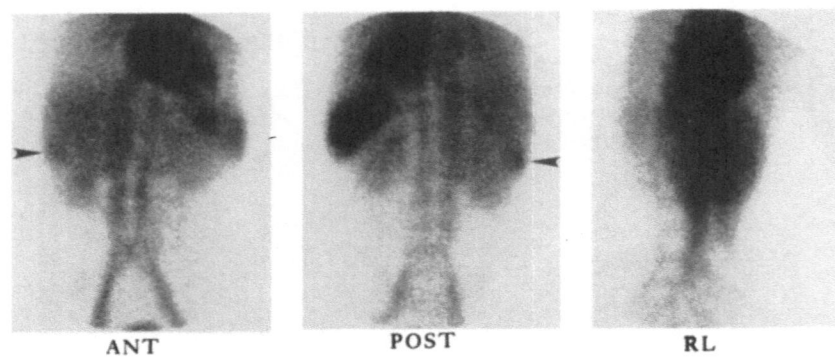

ANT   POST   RL   A

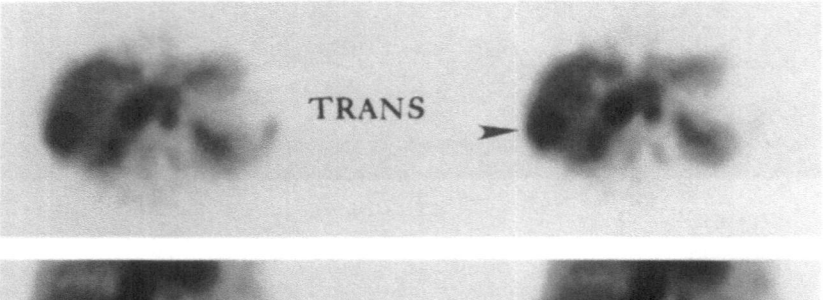

TRANS

COR

SAG

B

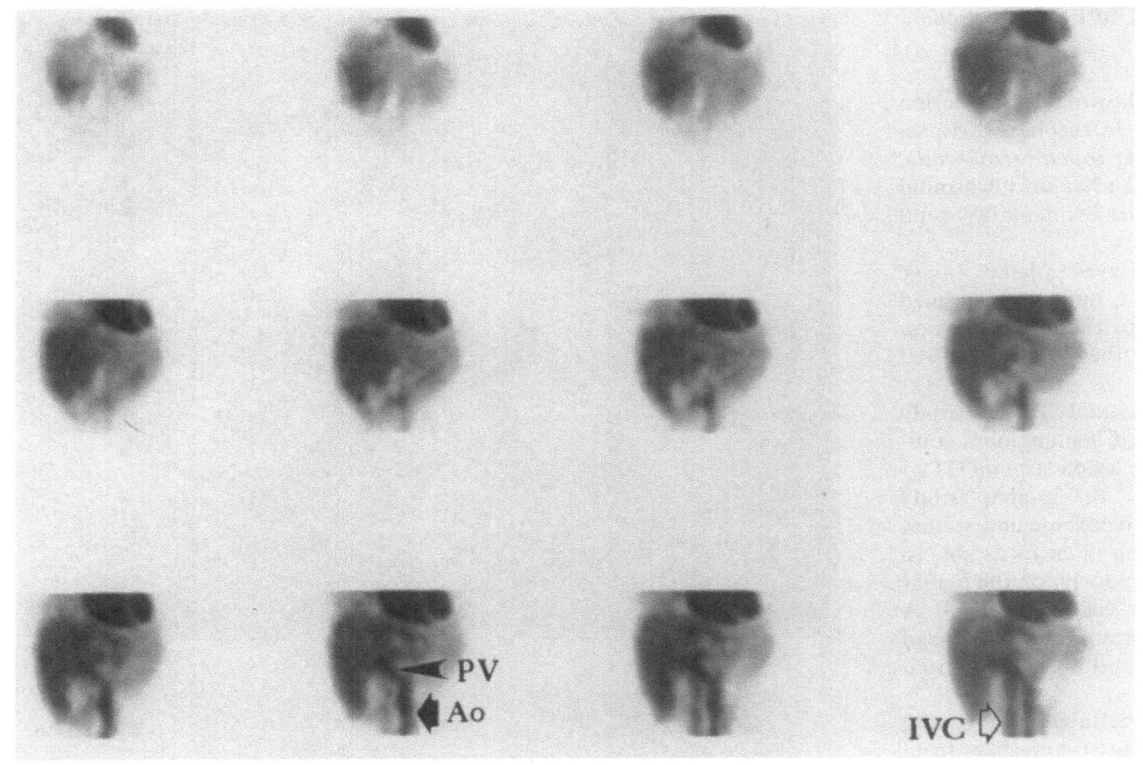

C

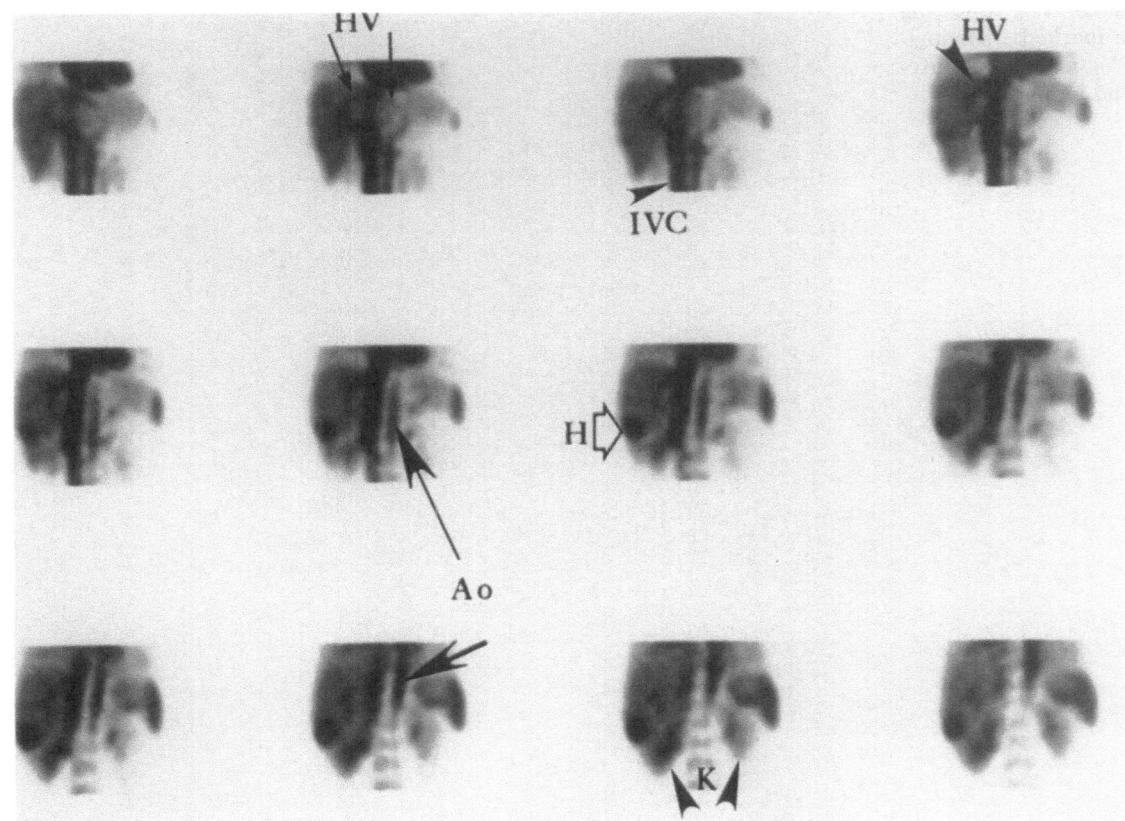

D

Figure 3.10. Value of SPECT over planar imaging.

A: Selected CT cut showing 1.5-cm lesion of uncertain etiology in the lateral aspect of left lobe of the liver (*open arrowhead*). The CT was obtained after an ultrasound study of the gallbladder incidentally found this lesion.

B: Anterior 1-hr delayed planar image shows mildly increased, but poorly defined uptake in the region of the lesion in question (*arrowhead*). Other views showed nothing.

C: The selected transaxial SPECT study shows a clearly defined hemangioma coinciding exactly with the lesion seen on CT.

Comment: Although the planar study might be read as positive, some uncertainty exists. This small lesion (2 cm) was able to be detected, although poorly, on the planar study because it was very superficial. A more central lesion would have blended into the liver background and likely would not have been seen.

D: Vascular anatomy defined on sequential transverse cuts. These are 6-mm slices from superior (*top left*) to inferior (*bottom right*) in the same patient (**A–C**). The following vascular structures are marked: hemangioma (H), portal vein (PV), aorta (Ao), inferior vena cava (IVC), and kidneys (K).

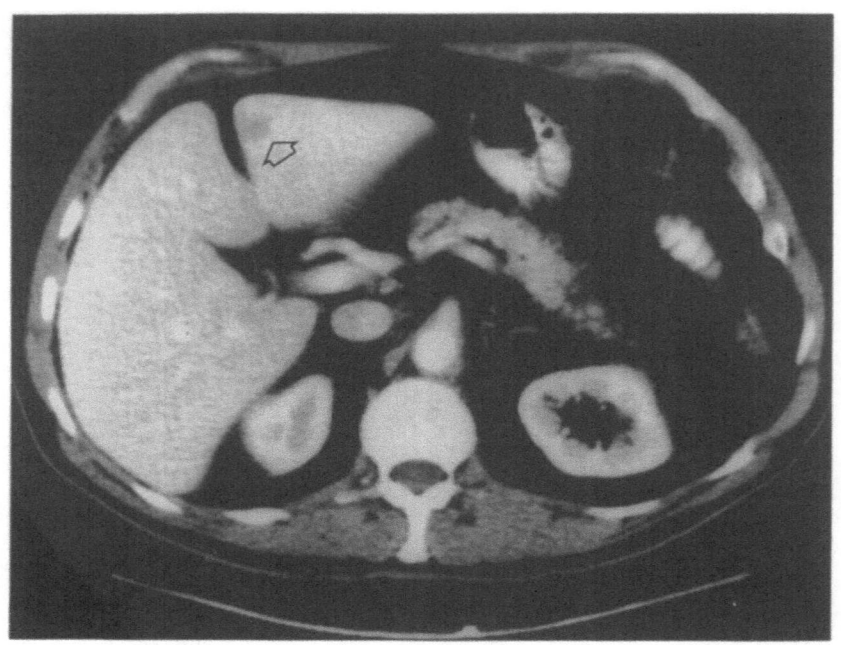

A

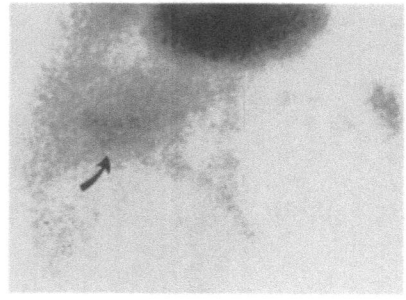

B

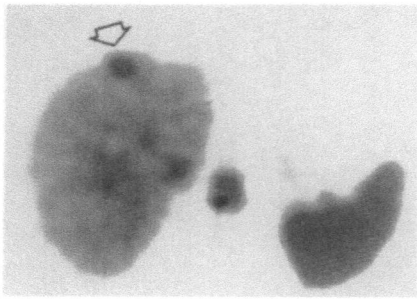

C

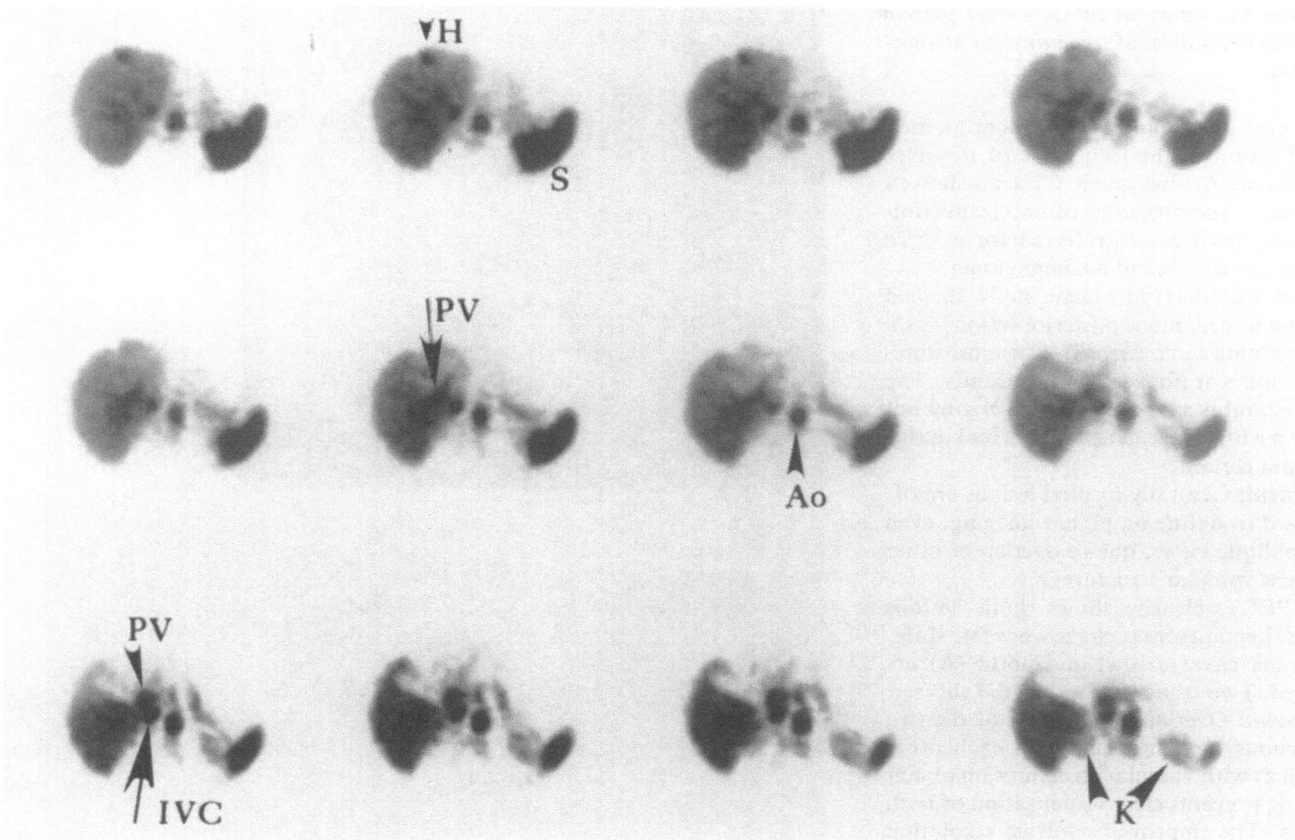

D

Figure 3.11. Value of SPECT over planar imaging: resolution of two adjacent heman-giomas.

A: During ultrasound of the kidney an inci-dental lesion in the right lobe of the liver was found. A subsequent CT scan showed two lesions (*arrowheads*) of uncertain etiol-ogy. The patient was referred for a ^{99m}Tc RBC study to rule out a hemangioma.

B: The 1-hr delayed planar study showed that the larger, more posterior lesion was a hemangioma (*arrowheads*). The more ante-rior lesion was not seen with certainty. The right lateral is suggestive (*white arrow*) but overlap with a vascular structure makes this far from certain.

Comment: Centrally located lesions are of-ten hard to define on planar imaging, even with oblique views, due to overlap of other adjacent vascular structures.

C: SPECT clearly shows both lesions to be hemangiomas (*arrowheads*). Infe-rior vena cava (*arrow*) and aorta (A) are marked. Two transverse sequential slices.

Comment: Overlapping activity of the two contiguous hemangiomas with each other as well as with vascular structures on planar imaging prevents clear visualization of both lesions. The improved contrast resolution and ability of SPECT to remove overlying activity is critical in this case. This example shows the clear advantage of SPECT over planar ^{99m}Tc RBC imaging.

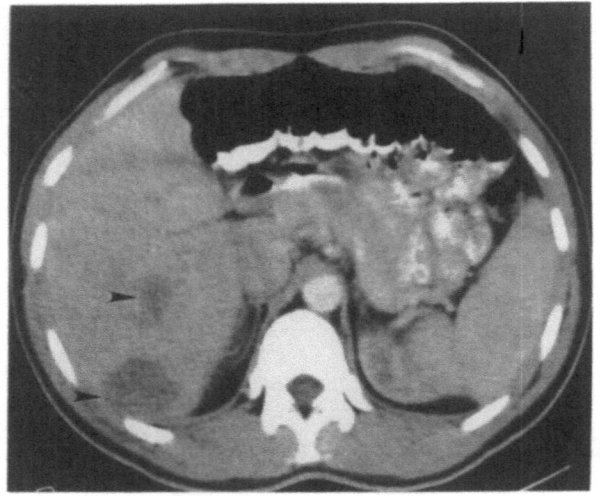

A

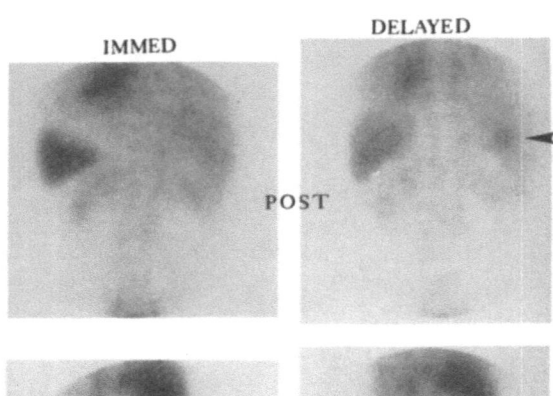

IMMED   DELAYED

POST

RL

B

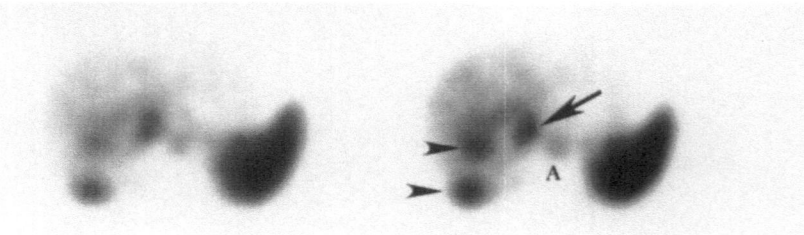

C

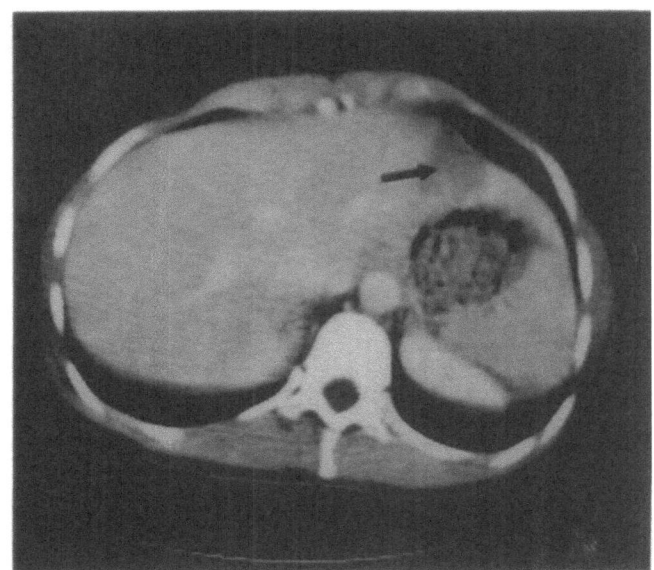

A

Figure 3.12. Value of SPECT over planar imaging for detecting a hemangioma adjacent to the spleen.

A: Contrast-enhanced CT scan shows lesion of uncertain etiology in the left lobe of the liver near the spleen and stomach.

B: A definite diagnosis cannot be made on the anterior and posterior planar views 1 hr postinjection because of overlapping activity of the hemangioma and spleen. The oblique views were equally unrewarding.

C: SPECT performed in 1984 using a single-headed gamma camera. The coronal (*left*) and transverse (*right*) selected slices clearly separate the hemangioma from the spleen. The diagnosis of hemangioma can be made on the SPECT study with confidence.

Comment: SPECT has clear superiority over planar imaging in the diagnosis of hemangiomas. Hemangiomas adjacent to the heart, spleen, kidney, and major vessels may be difficult to diagnose with planar imaging, but can usually be resolved with SPECT.

ANT   POST

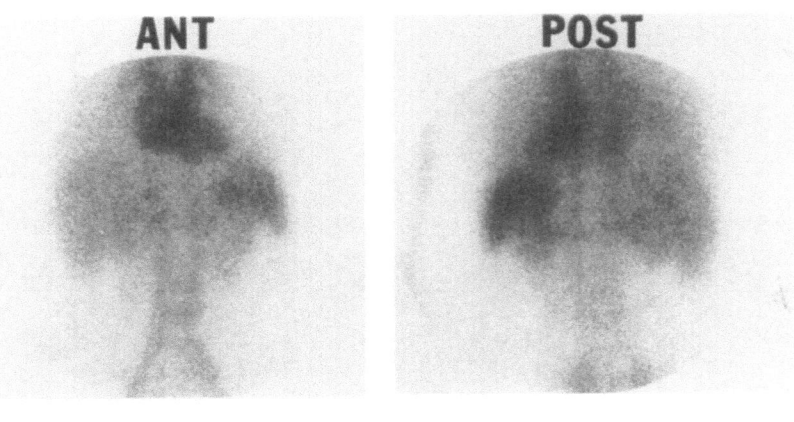

B

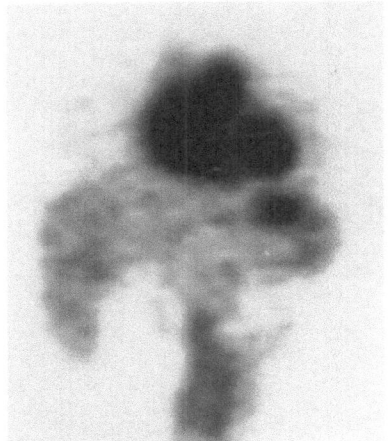
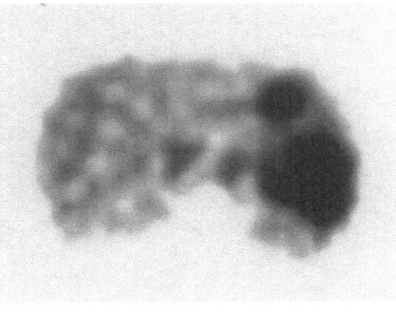

C

Figure 3.13. Potential for misinterpretation: difficulty in diagnosing liver adjacent to the heart on both CT and SPECT.

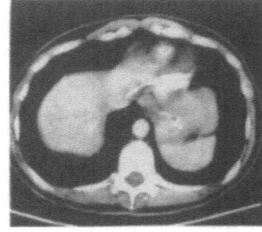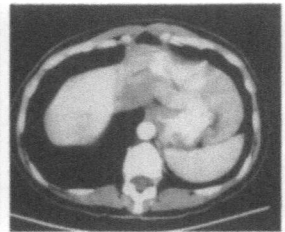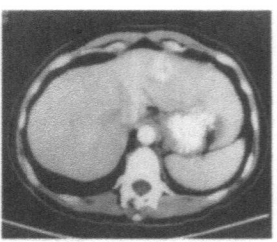

A

A: CT scan. There is 2.3-cm density with somewhat inhomogeneous contrast enhancement in the left lobe just inferior to the heart (note wax pencil marks on center CT slice pointing to the suspected lesion). The etiology of this lesion was uncertain and the possibility of artifact due to partial volume averaging of the cardiac apex was part of the differential diagnosis.

B: SPECT. Sequential, 6-mm transverse SPECT slices show a definite hemangioma in the left lobe inferior to the apex (A) of the heart. However, careful review of the images is necessary to avoid confusing the apex of the left ventricle with the hemangioma (H). The slice between A and H shows an apparent confluence of both due to partial volume averaging.

Comment: Appreciation of normal vascular anatomy is important in the interpretation of these studies. Reviewing sequential slices can help differentiate normal vascular structures from hemangiomas. This be-

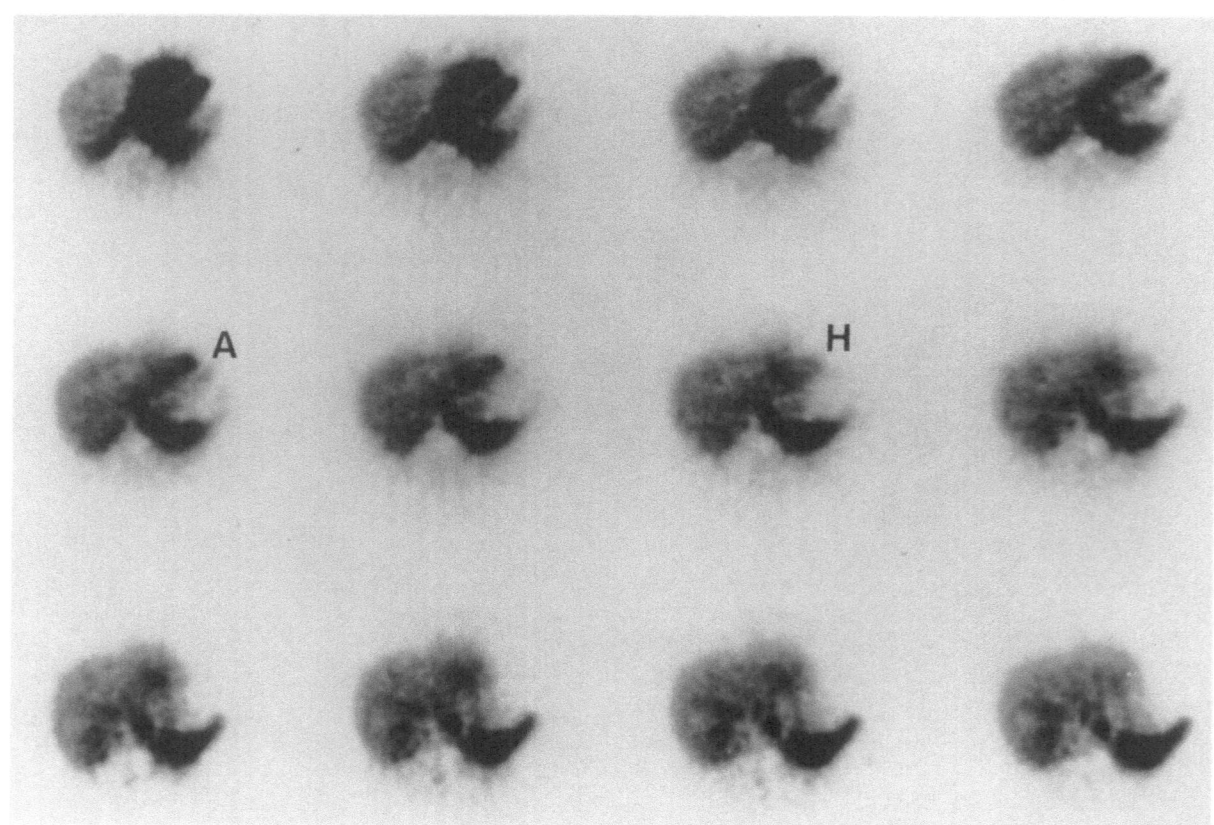

B

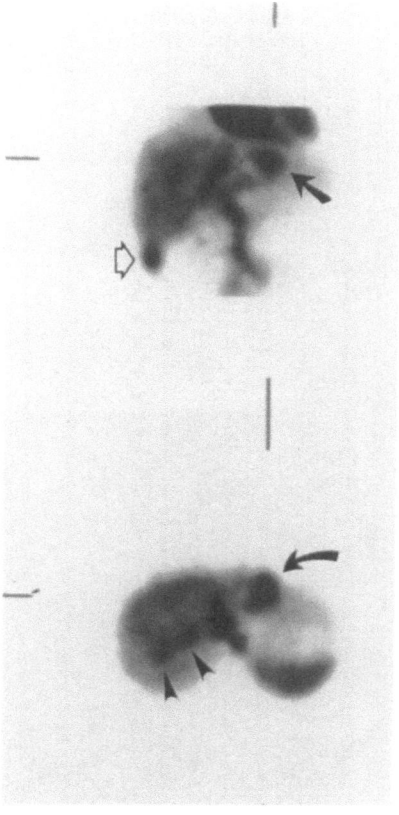

C

comes particularly important in the diagnosis of small lesions.

C: The coronal slice (*above*) clearly separates the left lobe hemangioma (*arrow*) from the left ventricle, also clearly seen on the selected transverse slice below (*arrow*). Note another hemangioma (*open arrowhead*) in the inferior aspect of the right lobe. This was not seen on CT. Two additional hemangiomas (*black arrowheads*) are suggested on this transverse slice (*below*), confirmed by reviewing adjacent transverse slices not shown here, and correlated with lesions seen on CT. This patient had five hemangiomas.

Comment: The coronal views can be particularly helpful in defining vascular structures and separating them from hemangiomas (see normal coronal vascular anatomy in Figure 3.9D).

Figure 3.14. Value of SPECT for visualizing more and smaller hemangiomas than planar imaging.

A: The ^{99m}Tc SC study (SC) shows a subtle negative defect (*arrow*) in the lateral aspect of the right lobe of the liver best appreciated when compared to the ^{99m}Tc RBC study. The ^{99m}Tc RBC immediate postflow study shows slightly increased uptake in the same area (*arrow*). The delayed RBC study shows definite uptake in the same location (*arrow*) consistent with a hemangioma in the posterior aspect of the right lobe.

B: Two cross-section transaxial SPECT slices demonstrate three hemangiomas (see corresponding CT slices in **C**). The transverse SPECT slice in the midright lobe (*left image*) shows the large lesion (*arrowhead*) seen on the planar study (**A**) posteriorly, but also shows a second smaller lesion at the same level more anteriorly (*small arrowhead*) not seen on the planar study. Note the adjacent kidney (**K**), which is clearly separate from the hemangioma. The other transverse slice (*right*) shows a small third hemangioma at the tip of the right lobe (*arrow*).

C: CT slices at comparable levels to the SPECT study (**B**). The largest lesion posteriorly measured 2.5 cm, the anterior lesion 1.8 cm, and the inferior lesion 1.1 cm.

Comment: The improved contrast resolution of SPECT allows detection of smaller and more centrally located hemangiomas than planar imaging.

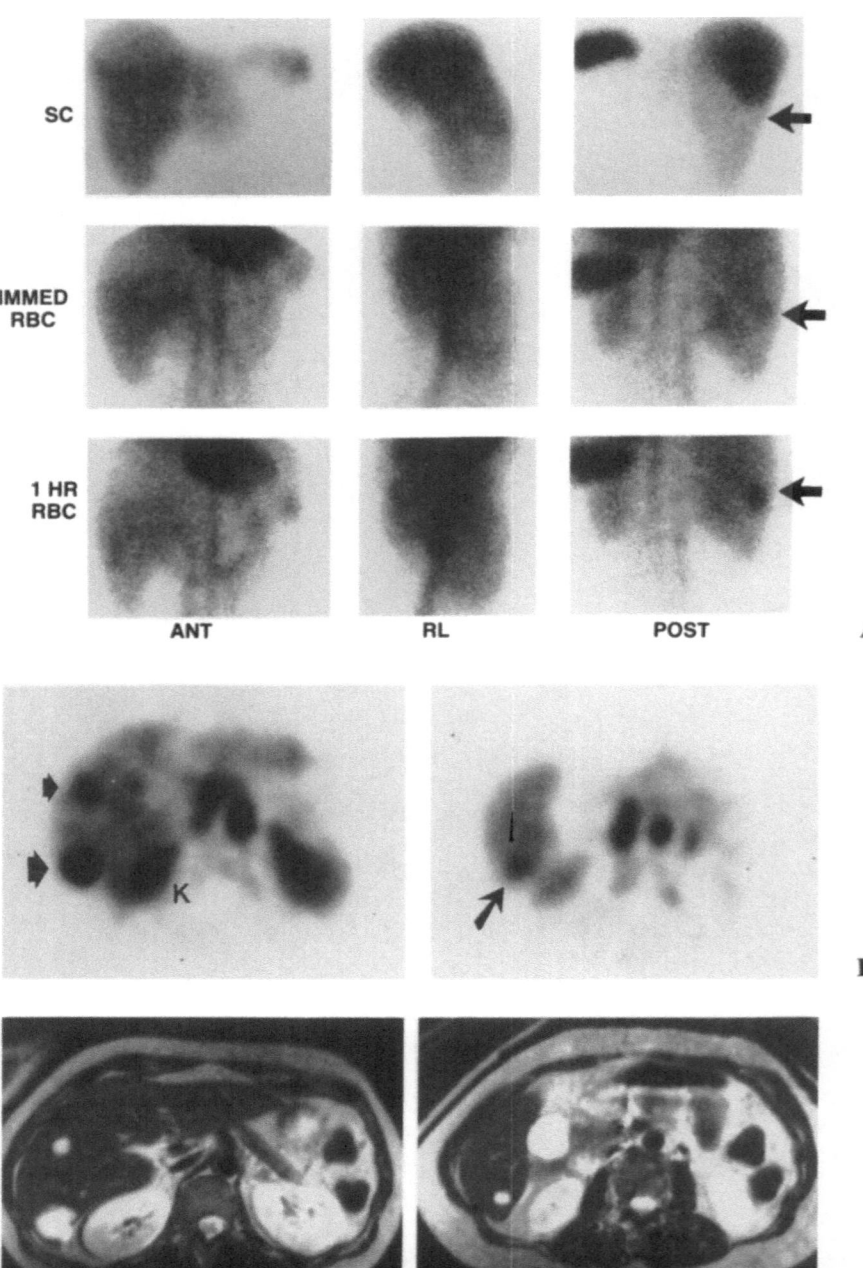

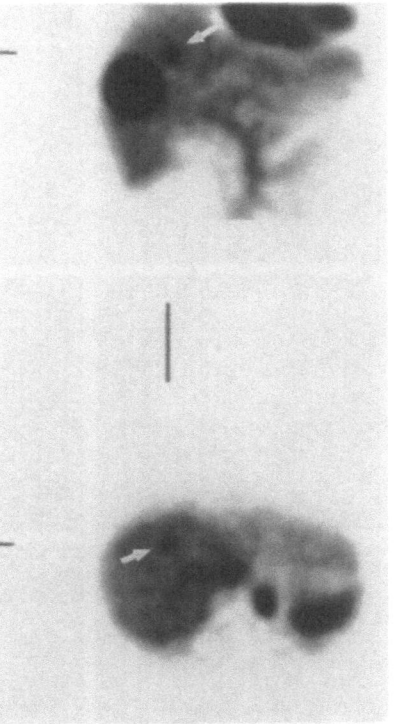

Figure 3.15. Value of triple-headed SPECT camera in comparison to a single-headed SPECT camera.

The SPECT coronal slice (*above*) clearly resolves the two adjacent hemangiomas, one large and the other small. The smaller lesion, seen also on the transverse slice (*arrow*), measured 9 mm on MRI. This study was performed with a three-headed dedicated high resolution SPECT system. Compare this with Figure 3.3D using a single-headed camera where only the large lesion was seen. This is the same patient.

Comment: ^{99m}Tc RBC scintigraphy may be falsely negative for lesions smaller than 2 cm in size using a single-headed SPECT camera or smaller than 1.4 cm in size using a triple-headed camera.

Figure 3.16. False positive ^{99m}Tc RBC studies.

A: Hepatoma. A–D: The flow study shows a hypervascular lesion in the left lobe of the liver. E: Increased uptake the on 1-min blood pool image (*arrow*). F: Further increased uptake is seen on the 2-hr delayed image.

Comment: Only five false positive ^{99m}Tc RBC studies have been reported: four hepatomas and one angiosarcoma.[14,17-19]

(Reprinted from Rabinowitz SA, McKusick KA, Strauss HW. ^{99m}Tc red blood cell scintigraphy in evaluating focal liver lesions. *Am J Roentgenol*. 1987;148:125–129. With permission.)

B-1: Hepatoma. The image labeled B-1 shows two sequential dynamic flow images (*above*) and 10-min postinjection static ^{99m}Tc RBC image (*below*) demonstrating a vascular lesion in the right lobe of the liver (*arrow*), nonuniform liver activity, and splenomegaly.

B-2: Sequential transverse SPECT images at 4 hr show focal increased uptake in the hepatoma (*arrow*).

Comment: Although hepatomas may rarely result in false positive studies as seen in Figures **A** and **B**, in fact most hepatomas have decreased uptake on delayed ^{99m}Tc RBC studies.[16]

(Reprinted from Itenzo C, Kim S, Madsen M, et al. Planar and SPECT ^{99m}Tc red blood cell imaging in hepatic cavernous hemangiomas and other hepatic lesions. *Clin Nucl Med*. 1988;13:237–240. With permission.)

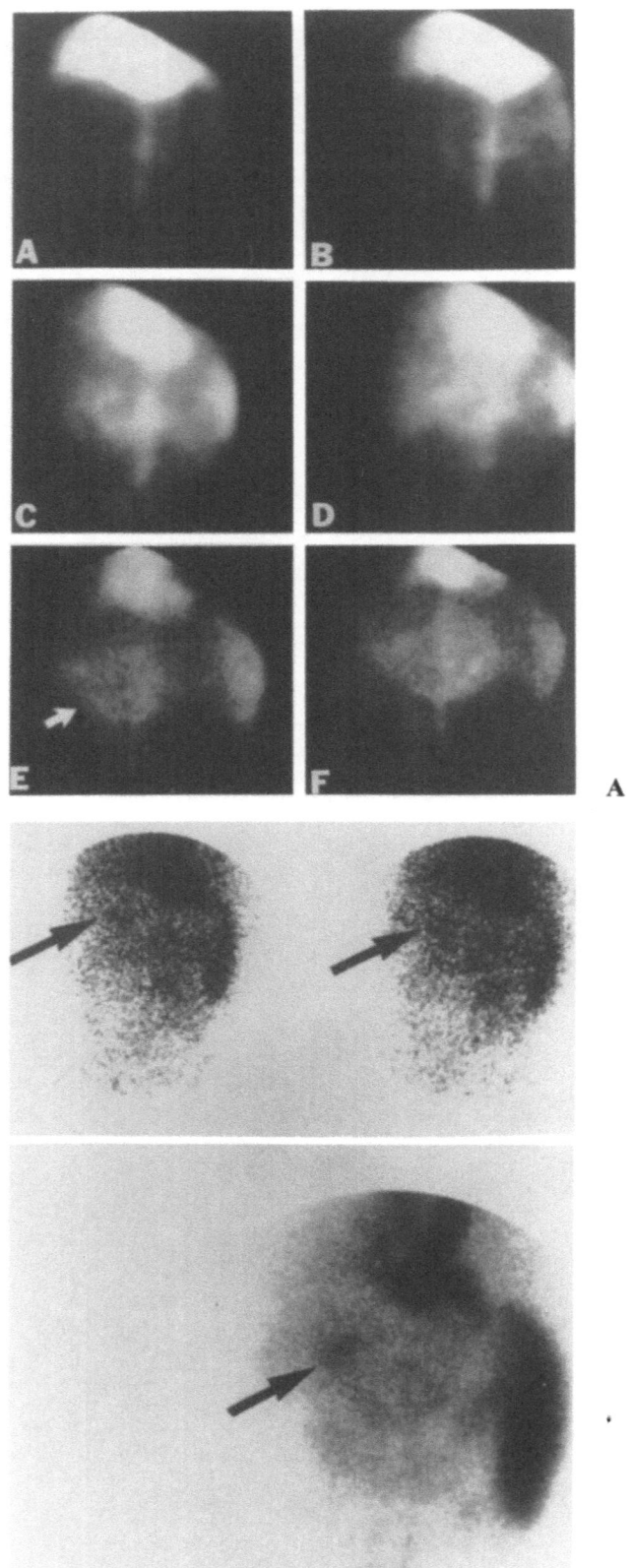

A

B-1

COMMAND: _

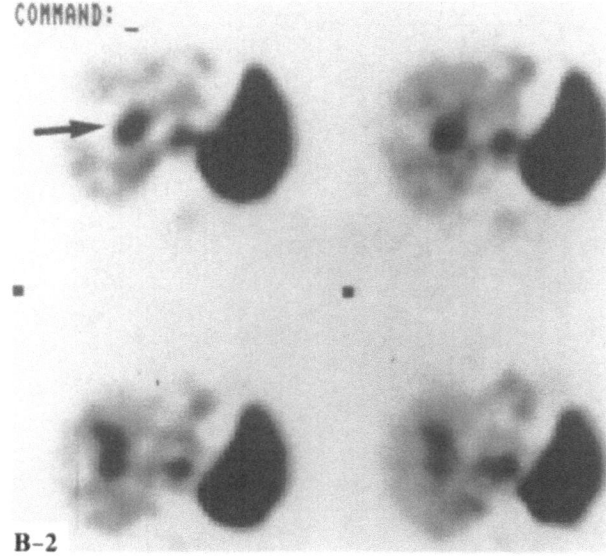

B-2

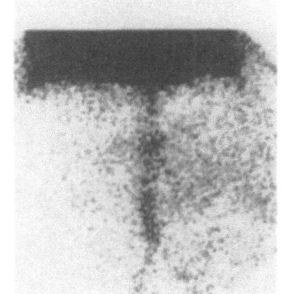

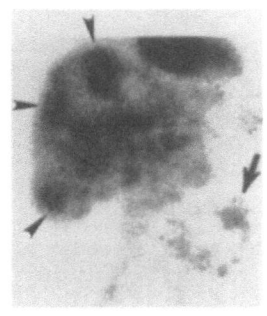

C

C: Hepatic angiosarcoma. Normal dynamic flow (*left*) and decreased uptake superiorly and laterally on the immediate blood pool image (*middle*) is seen. The delayed anterior view (*right*) at 3 hr shows multiple focal areas of increased uptake (*arrowheads*) that correspond to the initial cold areas. An extrahepatic site of increased uptake in the left abdomen is also noted (*arrow*). Contrast angiography revealed multiple areas of contrast puddling within the liver and also in the distal jejunum and proximal ileum consistent with hemangioma. Recurrent bleeding resulted in laparotomy and bowel resection. Pathological exam suggested malignant angiosarcoma. Subsequent uncontrollable bleeding resulted in the patient's death. Autopsy confirmed the diagnosis of angiosarcoma involving the liver, mesentery, and small bowel.

Comment: Angiosarcomas are rare malignant neoplasms of vascular origin. They have been associated with chronic exposure to thortrast, vinyl chloride, arsenicals, and radium and have also been associated with hemochromatosis.

(Reprinted from Ginsberg F, Slavin JD, Spencer RP. Hepatic angiosarcoma: mimicking of angioma on three-phase technetium-99m red blood cell scintigraphy. *J Nucl Med*. 1986;27:1861–1863. With permission.)

Figure 3.17. Hemangioendothelioma: typical findings in a 3-month-old infant.

A: Increased flow to the lesion (*arrow*) seen on 2-sec flow images.

B: The immediate postflow image shows heterogeneously increased uptake, particularly increased at the inferior border of the lesion. The 1-hr delayed image shows a similar pattern of uptake. The 2-hr delayed image is unchanged (patient positioning is somewhat different).

C: Comparable selected cross-section slices showing the patient's contrast-enhanced CT (*left*) and SPECT study (*right*). Both show the heterogeneity of the tumor with areas of increased (*arrows*) and decreased uptake on the SPECT study that correlate with CT.

Comment: This pattern of increased flow, increased uptake on immediate images, and increased uptake equivalent to heart and spleen on delayed images is diagnostic of hemangioendothelioma of childhood, although different from what is usually seen with adult hemangiomas. The extensive areas of degeneration and fibrosis may be a bit atypical.

Although these tumors are benign, they may simulate malignancy both clinically and on CT or angiography. They are usually symptomatic and typically present with hepatomegaly and congestive heart failure. Scintigraphy is diagnostic.[27]

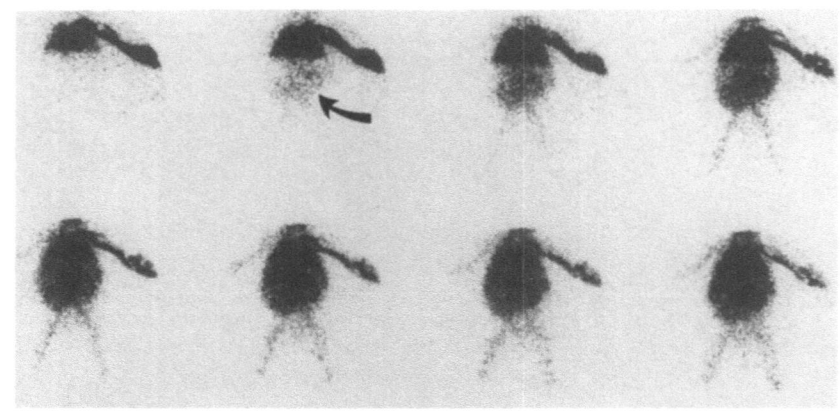

A

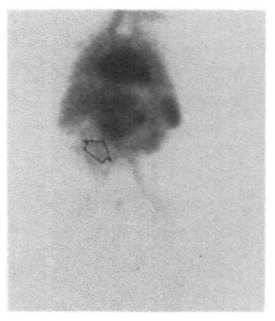 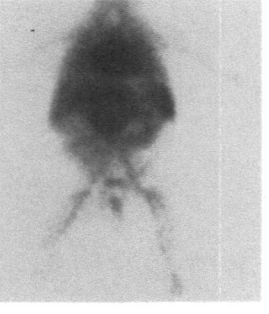 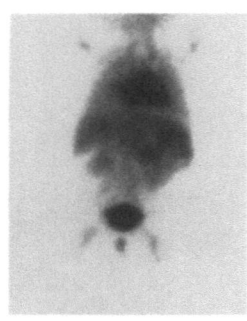

IMMEDIATE   1 HR   2 HR   B

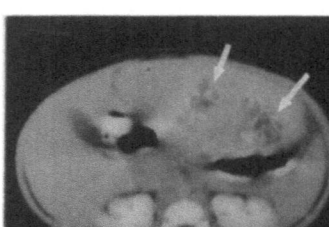 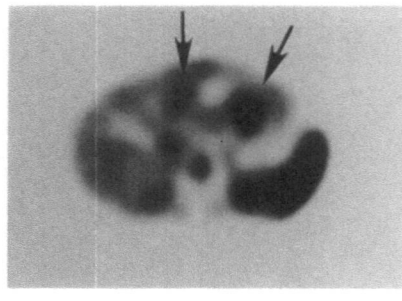

C

References

1. Ishak KG, Rabin L. Benign tumors of the liver. *Med Clin North Am.* 1975; 59:995–996.
2. Feldman M. Hemangioma of the liver. *Am J Clin Pathol.* 1958;29:160–162.
3. Edmondson HA. Tumors of the liver and intrahepatic bile ducts. In: Edmondson HA, *Atlas of Tumor Pathology.* Washington, DC: AFIP; 1958: 113–115.
4. McLoughlin MJ. Angiography in cavernous hemangioma of the liver. *Am J Roentgenol.* 1971;113:50–55.
5. Bree RL, Schwab RE, Glazer EM, Fink-Bennett D. The varied appearances of cavernous hemangiomas with sonography, computed tomography, magnetic resonance imaging and scintigraphy. *Radiographics.* 1987;7:1153–1175.
6. Itah Y, Ohnishi S, Ohtomo K, et al. Hepatic cavernous hemangioma in patients at high risk for liver cancer. *Acta Radiol.* 1987;28:697–701.
7. Taboury J, Porcel A, Tubiana JM, Monnier JP. Cavernous hemangiomas of the liver studied by ultrasonography. *Radiology.* 1983;149:781–785.
8. Freeny PC, Marks WM. Hepatic hemangioma: dynamic bolus CT. *Am J Roentgenol.* 1986;147:711–719.
9. Ashida C, Fishman EK, Zerhouni EA, et al. Computed tomography of hepatic cavernous hemangiomas. *J Comput Assist Tomogr.* 1987;11:455–460.
10. Stark DD, Felder RC, Wittenberg J, et al. Magnetic resonance imaging of cavernous hemangioma of the liver tissue-specific characterization. *Am J Roentgenol.* 1985;145:213–222.
11. Glazer GM, Aisen AM, Francis JR, et al. Hepatic cavernous hemangoma: magnetic resonance imaging. *Radiology.* 1985;155:417–420.
12. Ohtomo K, Itai Y, Yoshikawa K, Kokubo T, Iio M. Hepatocellular carcinoma and cavernous hemangioma: differentiation with MR imaging. Efficacy of T2 values at 0.35 and 1.5 T. *Radiology.* 1988;168:621–623.
13. Birnbaum BA, Weignreb JC, Megibow AJ, et al. Definitive diagnosis of hepatic hemangiomas: MR imaging versus Tc-99m labeled red blood cell SPECT. *Radiology.* 1990;176:95–101.
14. Rabinowitz SA, McKusick KA, Strauss HW. ^{99m}Tc red blood cell scintigraphy in evaluating focal liver lesions. *AJR.* 1984;143:63–68.
15. Brodsky RI, Friedman AC, Maurer AH, Radecki PD, Caroline DF. Hepatic cavernous hemangiomas: diagnosis with ^{99m}Tc-labeled red cells and single-photon emission CT. *Am J Roentgenol.* 1987;148:125–129.
16. Kudo M, Ikekub K, Yamamoto K, et al. Distinction between hemangioma of the liver and hepatocellular carcinoma: value of labeled rbc-SPECT scanning. *Am J Roentgenol.* 1989;152:977–983.
17. Drum DE. The radiocolloid liver scan in space-occupying disease. *Appl Radiol.* 1982;11:115–122.
18. Itenzo C, Kim S, Madsen M, et al. Planar and SPECT Tc-99m red blood cell imaging in hepatic cavernous hemangiomas and other hepatic lesions. *Clin Nucl Med.* 1988;13:237–240.
19. Ginsberg F, Slavin JD, Spencer RP. Hepatic angiosarcoma: mimicking of angioma on three-phase technetium-99m red blood cell scintigraphy. *J Nucl Med.* 1986;27:1861–183.
20. Tumeh SS, Benson C, Nagel JS, English RJ, Holman BL. Cavernous hemangioma of the liver: detection with single-photon emission computed tomography. *Radiology.* 1987;164:353–356.
21. Brunetti JC, Van Heertum RL, Yudd AP, Cooperman AM. The value of SPECT imaging in the diagnosis of hepatic hemangioma. *Clin Nucl Med.* 1988;13:800–804.
22. Brodsky RI, Friedman AC, Mauer AH, et al. Hepatic cavernous hemangio-

mas: diagnosis with ^{99m}Tc-labeled red cells and single-photon emission CT. *Am J Roentgenol.* 1987;148:125–129.

23. Keyes JW Jr, Harkness BA, Fahey FH, Ziessman HA. High-resolution SPECT of the liver and spleen. *Radiology.* 1989;173:215.

24. Ziessman HA, Silverman PM, Patterson J, et al. Improved detection of small cavernous hemangiomas of the liver with high-resolution SPECT. *J Nucl Med.* (in press.)

25. Brown RKJ, Gomes A, King W, et al. Hepatic hemangiomas: evaluation by magnetic resonance imaging and technetium-99m red blood cell scintigraphy. *J Nucl Med.* 1987;28:1683–1687.

26. Terriff BA, Gibney RG, Scudamore CH. Fatality from fine-needle aspiration biopsy of a hepatic hemangioma. *Am J Roengtenol.* 1990;154:203. (Letter to the Editor)

27. Kato M, Sugawara I, Okada A, et al. Hemangioma of the liver: diagnosis with combined use of laparoscopy and hepatic arteriography. *Am J Surg.* 1975;129:698–704.

28. Miller JH. Technetium-99m-labeled red blood cells in the evaluation of hemangiomas of the liver in infants and children. *J Nucl Med.* 1987;28:1412–1418.

29. Callahan RJ, Froelich JW, McKusick KA, et al. A modified method for the in-vivo labeling of red blood cells with Tc-99m: concise communication. *J Nucl Med.* 1982;23:315–318.

30. Atkins HL, Thomas SR, Buddemeyer U, Chervu LR. MIRD dose estimate report no. 14: radiation absorbed dose from technetium-99m labeled red blood cells. *J Nucl Med.* 1990;31:378–380.

31. Moinuddin M, Allison JR, Montgomery JH, Rockett JF, McMurray JM. Scintigraphic diagnosis of hepatic hemangioma: its role in the management of hepatic mass lesions. *Am J Roentgenol.* 1985;145:223–228.

32. Front D, Royal HD, Israel O, Parker JA, Kolodny GM. Scintigraphy of hepatic hemangiomas: the value of Tc-99m labeled red blood cells: concise communication. *J Nucl Med.* 1981;22:684–687.

33. Engel MA, Marks DS, Sandler MA, Shetty P. Differentiation of focal intrahepatic lesions with ^{99m}Tc-red blood cell imaging. *Radiology.* 1983;146:777–782.

34. Taylor RD, Anderson PM, Sinston MA, Blahd WH. Diagnosis of hepatic hemangoma using multiple radionuclide and ultrasound techniques. *J Nucl Med.* 1976;17:362–364.

35. Malik MH. Blood pool SPECT and planar imaging in hepatic hemangioma. *Clin Nucl Med.* 1987;12:543–545.

36. Brunetti JC, Van Heertum RL, Yudd AP, Cooperman AM. The value of SPECT imaging in the diagnosis of hepatic hemangioma. *Clin Nucl Med.* 1988;13:800–804.

37. Pettigrew RI, Witztum KF, Perkins GC, et al. Single photon emission computed tomographs of th liver: normal vascular intrahepatic structures. *Radiology.* 1984;150:219–223.

38. Sexton CC, Zeman RK. Correlation of computed tomography, sonography, and gross anatomy of the liver. *Am J Roentgenol.* 1983;141:711–718.

CHAPTER 4

Atlas of Hepatic Arterial Perfusion Scintigraphy

Harvey A. Ziessman

Colorectal carcinoma is the third most common cancer in the United States, accounting for 15% of malignancies. The liver is the most common site of metastasis. Approximately 25% of patients have hepatic metastases at the time of presentation and two-thirds eventually develop them. The presence of hepatic metastases has a major influence on prognosis. Survival in untreated patients with colorectal cancer metastatic to the liver varies from 1 to 22 months and is a function of the extent of hepatic disease at the time of presentation.[1] Complete surgical resection of liver metastases is the only curative therapy but is feasible for only a few patients who have solitary or unilobar metastasis. Conventional intravenous bolus chemotherapy (e.g., 5-fluorouracil) yields response rates of only 10% to 30%.[2]

Intraarterial administration of chemotherapeutic agents was first introduced in the early 1960s for the treatment of primary and metastatic tumors of the liver.[3-6] The rationale for this regional approach to chemotherapy is based on the differential blood flow to the tumor and normal liver. Hepatic metastases derive most of their blood supply from the hepatic artery, whereas normal liver cells are supplied predominantly by the portal circulation. Therefore, chemotherapy can be delivered preferentially to the tumor and systemic exposure, with potentially adverse side effects, is minimized.

Successful application of intraarterial chemotherapy requires that the drug be reliably delivered to the tumor-bearing area. This requires initial arteriographic assessment of the vascular supply of the tumor, followed by placement of a therapeutic catheter and then confirmation that the perfusion distribution from the catheter truly encompasses the entire tumor. Proper placement also minimizes perfusion of other organs.

The original technique of catheter placement for intraarterial chemotherapy involved percutaneous placement via the transfemoral or transaxillary approach. The catheters remained in place for several days or weeks depending on the length of the initial course of therapy. This short period of therapy is a major limitation of this technique. Repeat courses of therapy require placement of a new catheter. Early trials of this form of therapy reported a high incidence of catheter-related complications and an unpredictable or variable infusion pattern due to catheter movement. Modifications and improvements in this technique are still used at some centers.[7-9]

A more recent alternative approach is the surgical implantation of small-bore silastic catheters that are nonthrombogenic and can be used for long-term placement. The catheters are brought out through the skin and attached to an external delivery system or to an implanted subcutaneous slow-infusion pump system. The totally implanted drug delivery system surmounts many of the problems previously encountered with hepatic arterial infusion, including arterial occlusion, catheter thrombosis, vessel perforation, changes in flow distribution due to catheter movement, and the incidence of infection at the point where the percutaneous catheters enter the skin.[10,11] This approach is not without problems. It requires a major operation and the catheter cannot be repositioned once sutured in place. An experienced team approach is needed for successful intraarterial chemotherapy. This requires the coordinated efforts of the angiographer, surgeon, oncologist, and nuclear medicine physician.

Once the catheter is placed either percutaneously or surgically it is necessary to assess catheter placement. With percutaneous radiographic catheters, contrast arteriography can be performed. However, the high flow rates needed for good contrast angiography may not reflect the actual perfusion patterns that occur with the lower infusion rates used with chemotherapy delivery systems.[12,13] ^{99m}Tc macroaggregated albumin (^{99m}Tc-MAA) infused through the arterial catheter gives a more reliable estimation of blood flow distribution. Perfusion of the tumor-involved portion of the liver is essential for successful intraarterial chemotherapy. The patient's chemotherapeutic response can be predicted based on the adequacy of perfusion to the tumor on ^{99m}Tc-MAA hepatic arterial perfusion studies.[13,14] The presence of extrahepatic perfusion is a strong indicator of potential gastrointestinal and systemic toxicity. ^{99m}Tc-MAA hepatic arterial perfusion imaging has become the method of choice for making this assessment.[14-19]

Response rates of 34% to 72% have now been reported in a number of randomized studies of hepatic artery chemotherapy for the treatment of hepatic metastases from colorectal carcinoma.[20-25] There are ongoing investigations trying to improve intraarterial therapy with new chemotherapeutic agents. Fluorodeoxyuridine (FUDR) has been the most commonly used chemotherapeutic agent, because of its high hepatic extraction and short plasma half-life.[26-27] Adjunctive forms of therapy may play a role in improving the delivery and effectiveness of the primary chemotherapeutic agents, for example, leucovorin, bromodeoxuridine, starch microspheres, and vasoconstrictors.[20,28-33] Intraarterially administered radioactive microspheres (e.g., ^{90}Y) may also have an important place in future hepatic arterial therapy.[34-36]

Regional therapy has been used most commonly for colorectal carcinoma because it is the most common tumor metastatic to the liver; however, a variety of other primary and metastatic liver tumors have been treated with this form of therapy, including hepatocellular carcinoma, carcinoid, lung cancer, breast cancer, cholangiocarcinoma, and so forth. Reports of successful therapy of hepatocellular carcinoma with intraarterial Lipiodol are also encouraging.[37,38]

The clinical use of intraarterial chemotherapy has undergone many ups and downs in enthusiasm over the years. Since this form of therapy requires a concerted effort of experienced people to be most effective, it is probably best left to specialized centers. Nuclear medicine will continue to play an important diagnostic and perhaps even therapeutic role in regional intraararterial therapy as new methods of treatment become available.

The purpose of this atlas is to illustrate the technique for performing ^{99m}Tc-MAA hepatic arterial perfusion scintigraphy. Emphasis is placed on the interpretation of these studies.

Technique

The following technique is recommended for patients with *surgically implanted continuous infusion pumps.*[15]

Imaging Procedure

Patient preparation: No specific preparation necessary. A ^{99m}Tc sulfur colloid (^{99m}Tc SC) study performed 24 to 48 hr earlier is helpful for comparison with the ^{99m}Tc-MAA study.

Radiopharmaceutical administration: Agent: [99m]Tc macroaggregated albumin (^{99m}Tc-MAA). Dosage: 1 to 4 mCi (37–148 MBq) for planar imaging and 5 to 6 mCi (185–222 MBq) for SPECT, injected in a small fluid volume of 0.5 to 1.0 cc. Route: Intraarterial. Technique for injection into subcutaneously implanted infusion pump: A 22-gauge 1″ Huber needle is inserted into the sideport of the infusion pump. Free flow is acertained. Then the ^{99m}Tc-MAA is slowly and continuously injected over 1 to 2 min and flushed with 10 cc saline. Before the needle is removed, 5 cc of heparin (10 units/cc) is infused.

Image acquisition: Camera: Large field of view. Collimator: low energy all-purpose, parallel-hole collimator. Energy window: 15% window centered over a 140-keV photopeak. Positioning: All images acquired supine. Views: 500,000 count anterior image, then posterior, right lateral, left lateral, and anterior chest views for equal time.

If extrahepatic gastric perfusion suspected: 4 g of sodium bicarbonate–citric acid–simethicone effervescent granules (EZ-Gas, Sparkles) should be administered in 100 cc of water.[39] The patient must be encouraged not to eruct. Repeat anterior and left lateral images are then repeated.

Alternative technique for injection into a *percutaneously placed catheter*[14]:
1. A three-way stop cock is placed as close as possible to the site of catheter entry into either the brachial or femoral artery.
2. With the patient positioned under a large field of view gamma camera, the catheter is gently flushed with 10 to 20 ml of normal saline.
3. ^{99m}Tc-MAA, 2 to 4 mCi (74–148 MBq) in 0.2 cc volume is introduced via the three-way stop cock into the hepatic artery catheter.
4. Immediately the flow of the external pump is advanced to a high rate (200 ml/hr).
5. The progress of the radioactive injectate is visually monitored on the camera persistence scope as it migrates from the external aspect of the catheter to the origin of the hepatic artery. As the bolus approaches the liver, the flow rate on the external pump is decreased to the exact rate at which the chemotherapy is to be delivered (generally between 10 and 21 ml/hr).

6. The radioaggregate distribution within the liver is again continuously monitored on the persistence scope until sufficient injectate resides within the liver to allow multiple 500,000-count views of that organ, which can be obtained with 100 to 300 sec/view.

Physiological Mechanism of the Radiopharmaceutical

The ^{99m}Tc-MAA has a particle size of 10 to 90 μm. When injected into the hepatic artery it distributes according to blood flow and will be trapped on first pass in the arteriolar capillary bed of the liver. The resulting image is a map of blood flow distribution. The irregularly shaped and malleable ^{99m}Tc-MAA injected particles occlude a small percentage of the arteriolar capillary bed of the liver and begin to break down into smaller and smaller ^{99m}Tc-MAA particles (effective half-life of about 3 hr), which are subsequently taken up by the reticuloendothelial system.

Arteriovenous shunting through the liver or tumor results in trapping of a portion of these particles within the lung. Of course chemotherapy would not be trapped in the lung, but this gives an indication of the amount of drug that would not be delivered to the tumor but rather to sensitive dose-limiting tissue, such as gastrointestinal epithelium and bone marrow.

Extrahepatic perfusion is manifested by uptake within visceral organs, most commonly the stomach, spleen, and bowel due to undesirable blood flow directed away from the liver by other local vessels. This may results from improper positioning of the catheter, difficulty in optimally positioning the catheter due to the patient's variant anatomy or thrombosis. Since the takeoff of several vessels of the hepatic vasculature may be close to each other (e.g., the proper hepatic, right and left hepatic and left gastric), the direction of the blood flow can be very dependent on the rate of delivery. The high flow rates needed for good contrast angiography may not reflect the actual perfusion patterns that occur with the lower infusion rates used with chemotherapy delivery systems.[12,13] The distribution of radiotracer injected at flow rates of 1 ml/min has been shown to be quite different from rapid 15-ml bolus injections. In fact, there are significant differences in organ visualization with contrast angiography performed at 4 ml/sec versus 1 ml/sec.[12] This discrepency is due to turbulent and streamlined flow and by local increases in pressure, which can even reverse the direction of blood flow. Therefore, blood flow assessed for large vessels and the organ perfusion pattern at the capillary level may be quite different. Since chemotherapeutic agents gain access to the tumors at the capillary level, the radionuclide study is ideal for this evaluation.

Estimation of Radiation Absorbed Dose

Separate calculations are shown for:

1. good liver perfusion, mild A-V shunting to the lung, and no extrahepatic perfusion

2. decreased hepatic uptake (40%), significant extrahepatic perfusion (40%), and some A-V shunting (10%).

Dose calculations assume a mean effective half-life of radioparticles in the liver of 4 hr. Intestinal extrahepatic perfusion was calculated for tissue volumes of 10, 40, and 100 g, since extrahepatic perfusion is usually not diffuse throughout the abdomen, but localized. The modififed dose estimates are adopted from Croft[40] and Thrall.[41]

Table 4.1. Estimated radiation absorbed dose.

	Rads/mCi
Hepatic perfusion (96%):	
Extrahepatic perfusion (0%)	
A-V shunting to lung (4%)	
Liver	0.26
Lungs	0.01
Intestines	
10 g	0.0
40 g	0.0
100 g	0.0
Total body	0.02
Gonads	0.02
Hepatic perfusion (40%):	
Extrahepatic perfusion (40%)	
A-V shunting to lung (10%)	
Liver	0.11
Lungs	0.06
Intestines	
10 g	0.10
40 g	0.03
100 g	0.01

Visual Description and Interpretation

The effectiveness of intraarterial chemotherapy will be maximized if there is perfusion of the entire tumor-involved liver. Side effects will be minimized when there is no extraphepatic perfusion (e.g., to the stomach, spleen, and bowel). To ensure optimal placement, ^{99m}Tc-MAA hepatic arterial perfusion studies should be performed after the initial placement of the catheter and before the initiation of chemotherapy. Once proper placement and optimal perfusion is assured, chemotherapy can commence. If the patient develops symptoms of drug toxicity or has a poor response to chemotherapy, the study should be repeated to determine if the perfusion pattern has changed.

This atlas illustrates and describes techniques recommended for performing and interpreting ^{99m}Tc-MAA hepatic arterial perfusion studies. Emphasis is placed on ways to maximize the diagnostic information available from these studies to aid the clinician in providing effective intraarterial chemotherapy with minimal side effects.

Acknowledgments. The author wishes to acknowledge the contributions to this chapter by James H. Thrall, John W. Gyves, William D. Ensminger, Jack E. Juni, Richard L. Wahl, David M. Williams, John W. Keyes, Jr., and the many others at the University of Michigan where much of this work was initiated and performed.

Atlas Section

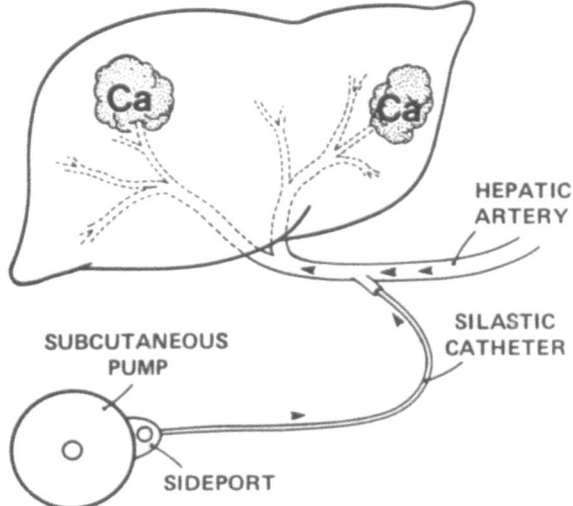

HEPATIC
ARTERY

SILASTIC
CATHETER

SUBCUTANEOUS
PUMP

SIDEPORT

Figure 4.1. Schematic diagram of a totally implanted drug delivery system with catheter inserted so as to feed drug to entire tumor-bearing liver.

(Reprinted from Ensminger W, Niederhuber J, Dakhil S, Thrall J, Wheeler R. Totally implanted drug delivery system for hepatic arterial chemotherapy. *Cancer Treat Rep*. 1981;65:393–400. With permission.)

Figure 4.2. Surgically implanted Infusaid pump.

A: The Infusaid pump (Infusaid Corp., Norwood, MA) measures 7 × 9 cm and weighs 200 g. It is implanted subcutaneously in a pocket created in the abdominal wall.

B: Cross-section of the infusion pump, showing the two chambers. The outer chamber contains a fluorocarbon liquid in equilibrium with its vapor phase. At 37° C, the vapor pressure is about 300 mm Hg greater than atmospheric pressure, providing a continuous power source and constant rate of delivery (usually 3–6 ml/day). The inner chamber, which contains the drug, has a 50-ml capacity and is refilled every 10 to 14 days by percutaneous injection. The side port bypasses the pumping mechanism and can be used for direct injection of drugs best given by bolus or short infusion or to inject radionuclides for perfusion imaging. The hepatic arterial catheter is made of polymeric silicone and has an inner diameter of 0.63 mm (Silastic, Dow Corning).

(Reprinted from Ziessman HA, Thrall JH, Yang PJ, et al. Hepatic arterial perfusion scintigraphy with Tc-99m MAA. *Radiology*. 1984;152:167. With permission.)

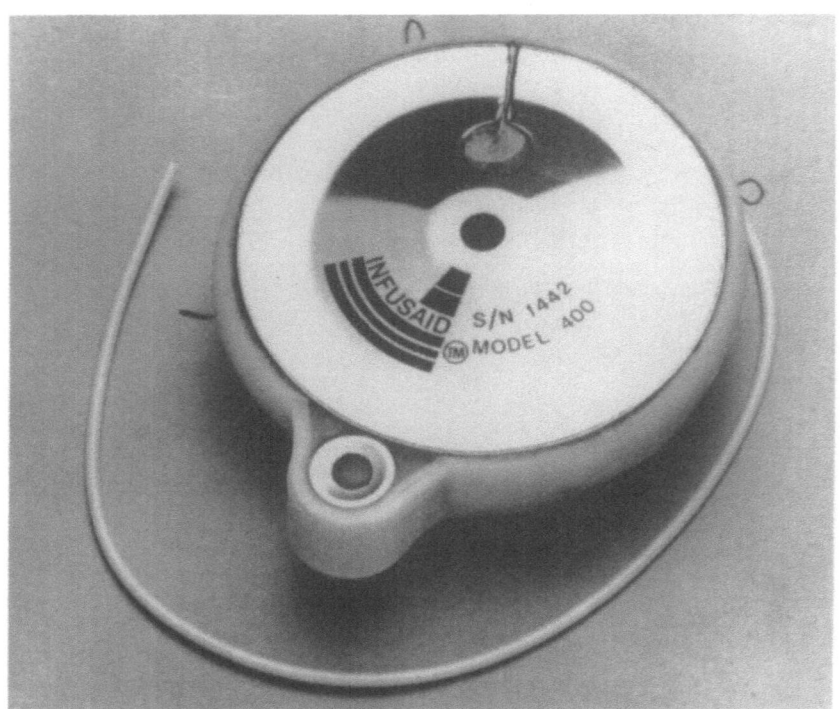

A

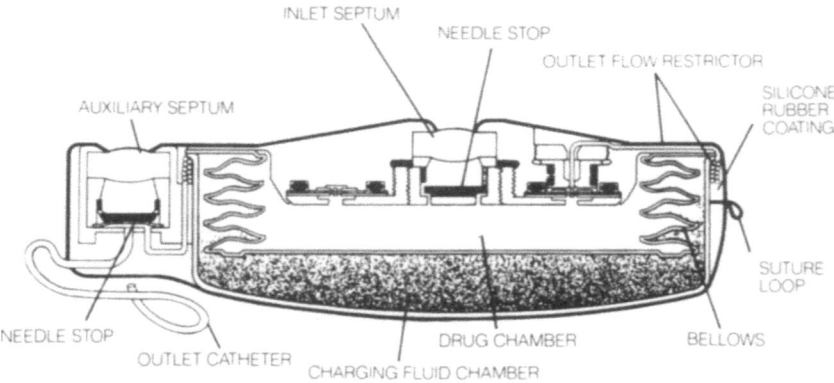

B

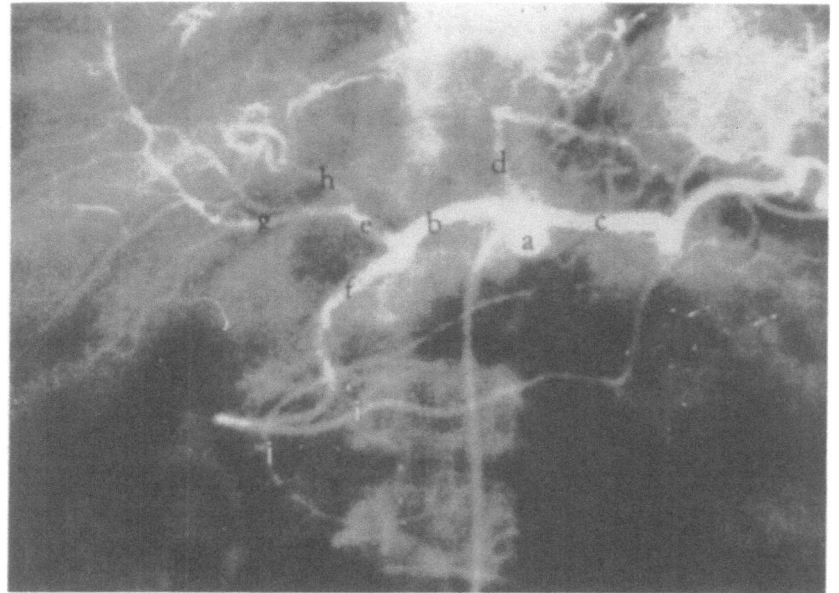

Figure 4.3. Standard hepatic vascular anatomy.

The common hepatic artery (b) originates from the celiac axis (a), as does the splenic artery (c) and left gastric (d). The gastroduodenal artery (f) branches off the common hepatic artery, which then becomes the proper hepatic artery (e), then dividing into the right (g) and left hepatic artery (h).

Comment: This normal hepatic vascular pattern is reported to occur in only 55% of persons; 45% of person have so-called variant anatomy.[42] The list of possible variants is large; two of the most common are the right hepatic replaced to the superior mesenteric and the left hepatic arising from the left gastric (see Figures 4.4C,D). Proper placement of the intraarterial catheter requires initial assessment of the patient's vascular anatomy.

Figure 4.4. Surgical approach to placement of intraarterial catheters.

A: Catheter placement in a patient with standard hepatic arterial anatomy. The gastroduodenal artery is ligated and the catheter is placed at the junction of gastroduodenal artery and hepatic artery. The right gastric artery has been ligated. A close-up view of the beaded catheter is shown in the inset. The Silastic catheter outer diameter is 2.3 mm and the inner diameter is 0.63 mm.

B: The right and left hepatic arteries originate too close to the gastroduodenal artery to permit mixing of drug and equal distribution to all areas of the liver. This is a variation of standard anatomy. In this situation the gastroduodenal artery and the right gastric artery are ligated. The splenic artery is ligated and the catheter positioned at the junction of the splenic artery and the celiac axis.

C: The left hepatic artery is depicted arising from the left gastric artery. The middle hepatic artery and the right hepatic artery originate in standard fashion from the proper hepatic artery. Total liver perfusion requires two catheters. One catheter is placed in the ligated left gastric artery and positioned at the junction of the left hepatic artery (*arrowhead*). The diameter of these peripheral vessels is small and care is taken not to occlude the lumen by the catheter. The right lobe is perfused by a catheter placed in the gastroduodenal artery.

D: The presence of a replaced right hepatic artery originating from the superior mesenteric artery requires use of two catheters. A special catheter is used to gain access to the replaced right hepatic artery (see *inset*). A small arteriotomy is fashioned in the lateral wall of the replaced right hepatic artery. A 6–0 cardiovascular "purse string" stitch is placed around the 2-mm arteriotomy. A narrow tip catheter is inserted so the thicker portion of the catheter abuts on the arterial wall. A second catheter is placed in the gastroduodenal or splenic artery, dedending on the anatomy.

(Reprinted from Niederhuber JE, Ensminger WD. Surgical considerations in the management of hepatic neoplasia. *Semin Oncol*. 1983;10:135–147. With permission.)

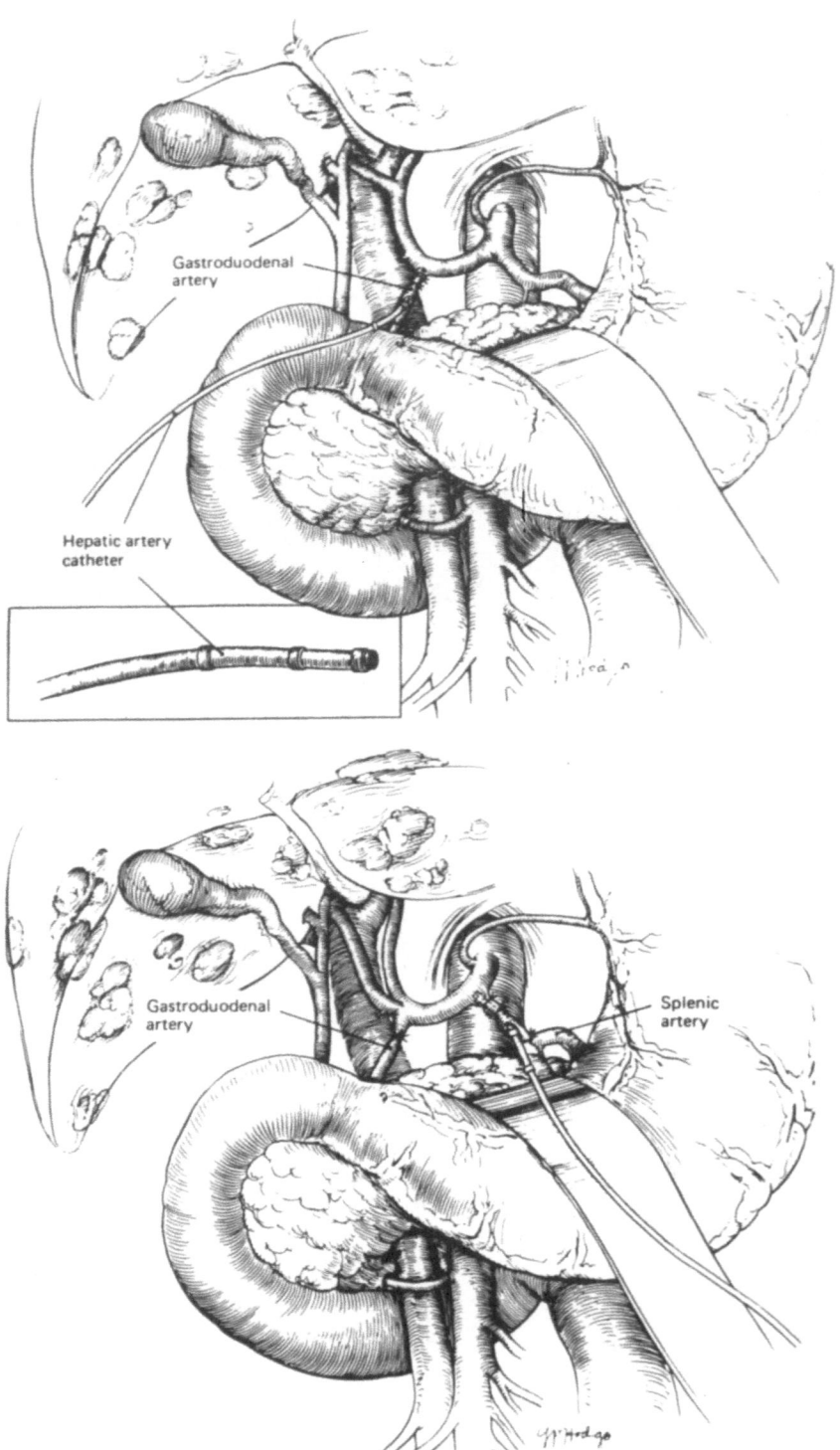

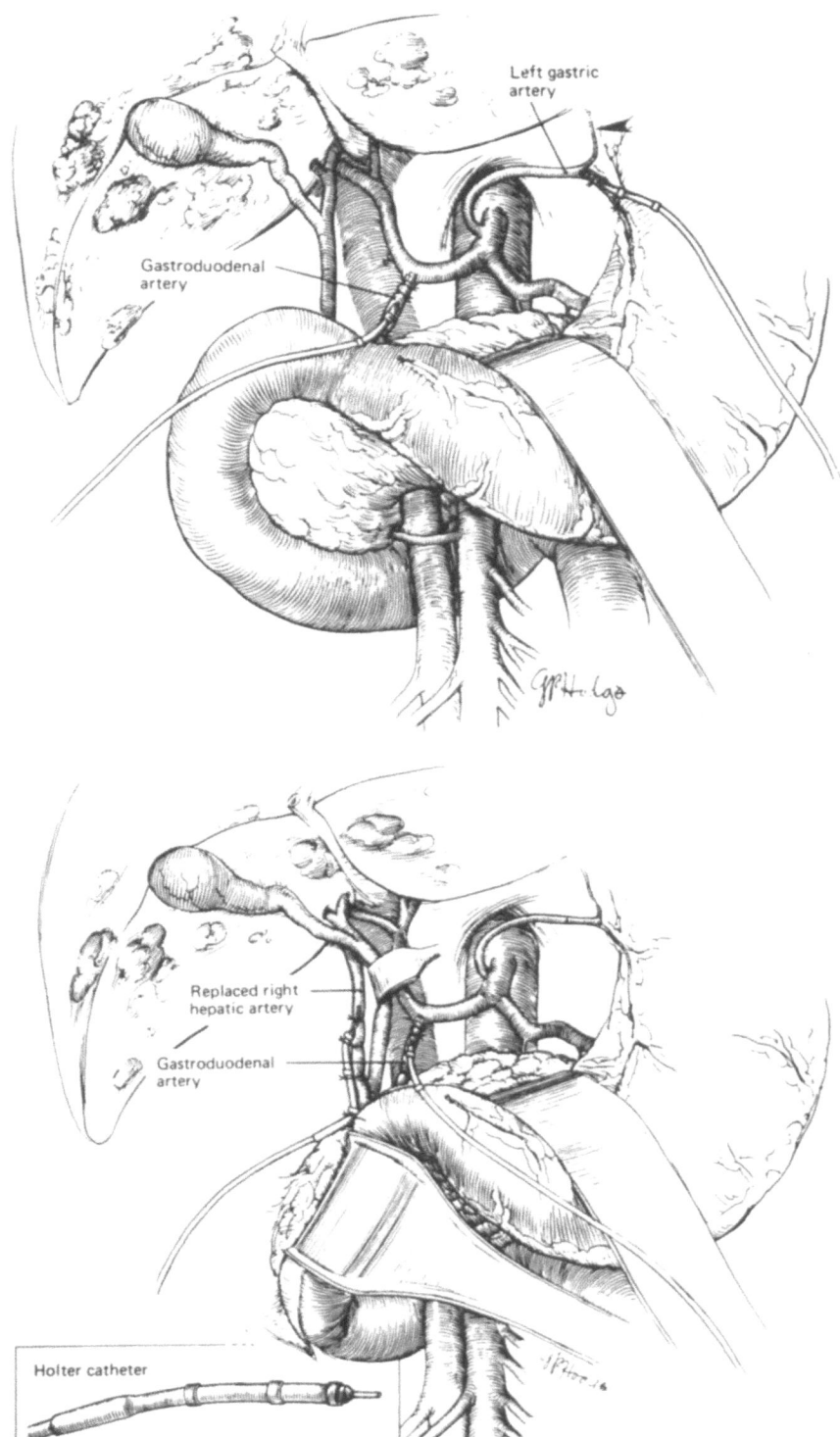

C

D

Figure 4.5. Clinical and Tc SC scintigraphic improvement with intraarterial chemotherapy.

A: Improvement with a combination of surgery and intraarterial chemotherapy. The bulk of a large tumor mass (hepatoma) in the right lobe of the liver was resected at surgery and an intraarterial catheter placed for chemotherapy of the residual tumor. Follow-up with ^{99m}Tc SC liver–spleen scans over 2 years' time show continuing improvement (*from left to right*, with first study *top left* and most recent study *bottom right*).

B: Successful intraarterial chemotherapy in a patient with colon cancer metastatic to the liver. The prechemotherapy ^{99m}Tc SC study (*above*) shows extensive metastases. The follow-up ^{99m}Tc SC scan 1 year later (*below*) shows marked improvement.

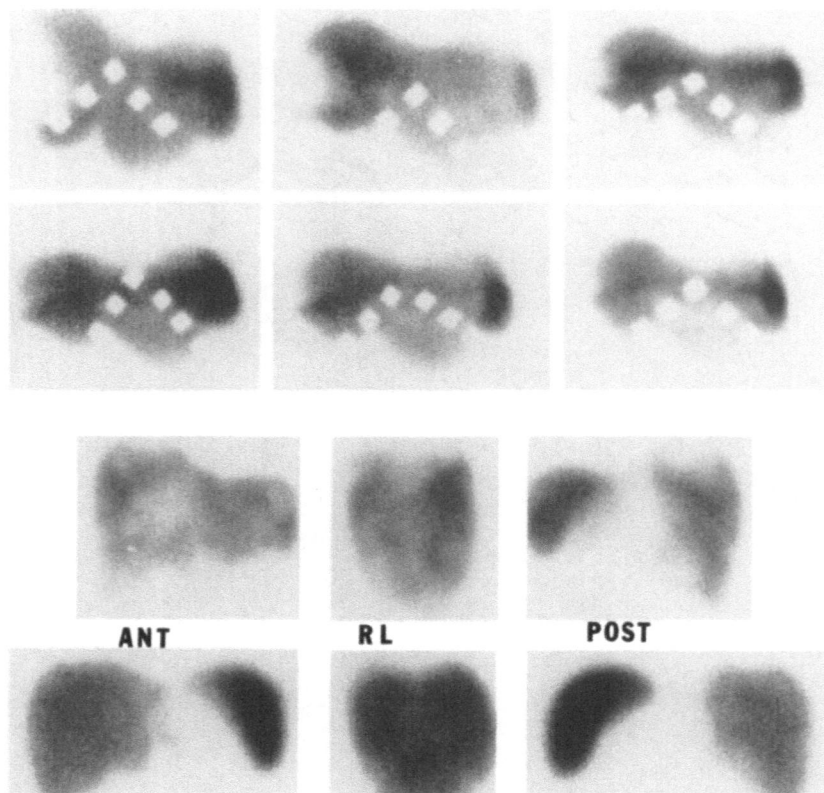

ANT   R L   POST

A

B

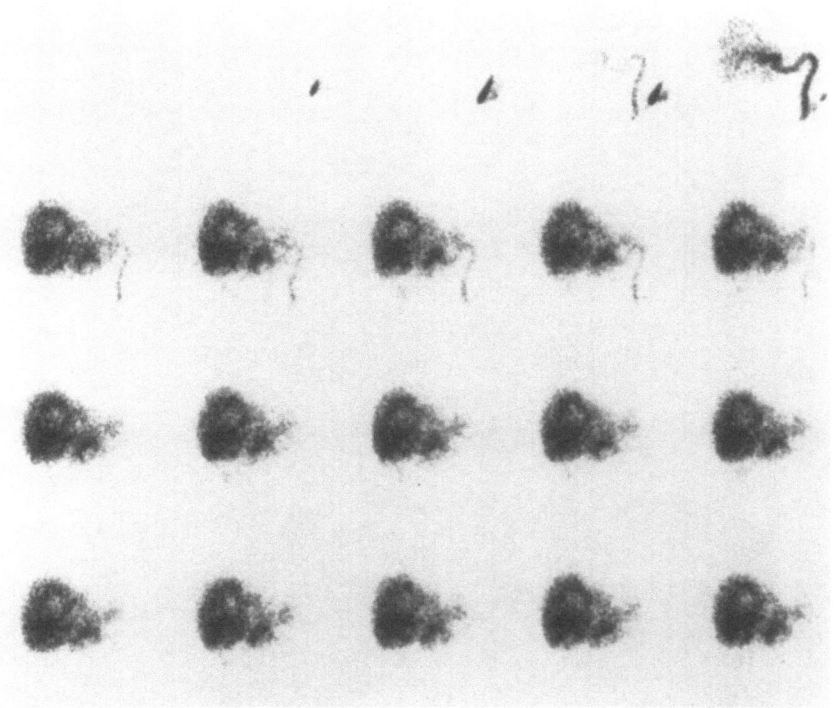

A

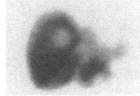

B

Figure 4.6. Intraarterial radionuclide angiography versus static imaging.

A: Two-second per frame ^{99m}Tc-MAA blood flow images. The catheter is not obstructed and there is good flow to entire liver. However, the images are low resolution and in only one view. Since the radiotracer should be infused slowly over 60 sec using a surgically implanted catheter and given at the rate of chemotherapy administration for percutaneously placed catheters, flow images are impractical and unnecessary in most cases.

B: Static high count anterior image. This 500,000-count image acquired soon after injection shows a similar distribution of blood flow to the liver as the flow images in **A**. Static imaging postinjection gives the same information with improved resolution and also allows imaging in multiple views.

Figure 4.7. Intraoperative ^{99m}Tc-MAA study using computer subtraction techniques to ensure proper catheter placement.

The surgeon manipulated the catheter until optimal perfusion was obtained (**D**). Computer subtraction techniques were used to remove previous injection radioactivity. The first two catheter positions perfused only the right lobe of the liver (**A,B**), whereas the third attempt resulted in perfusion to the left lobe and perhaps stomach (**C**).

Comment: Placing an intraarterial catheter surgically to perfuse the entire tumor-bearing liver takes considerable experience. Once the catheter is fixed, it cannot be removed. Therefore, a surgeon just learning this technique would be advised to request an intraoperative radionuclide study to ensure proper placement.

(Reprinted from Thrall JH, Gyves JW, Ziessman HA, Ensminger WD. Hepatic arterial perfusion studies. In: Thrall JH, Swanson DP. *Diagnostic Interventions in Nuclear Medicine.* Chicago: Yearbook Medical Publishers; 1985:176–194. With permission.)

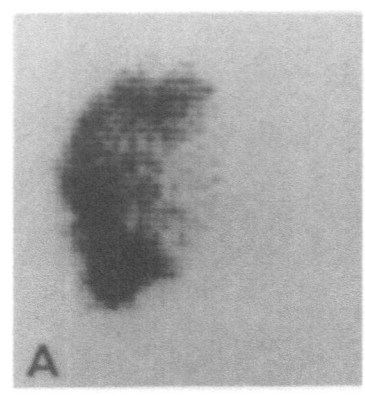
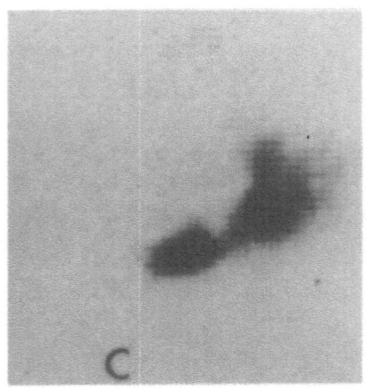
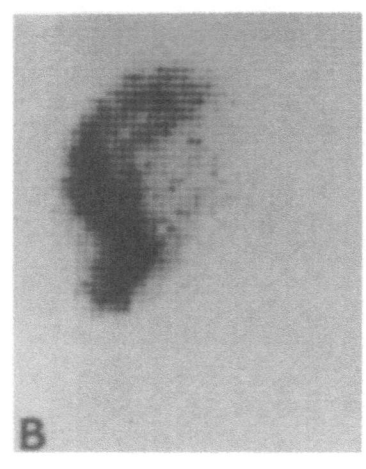
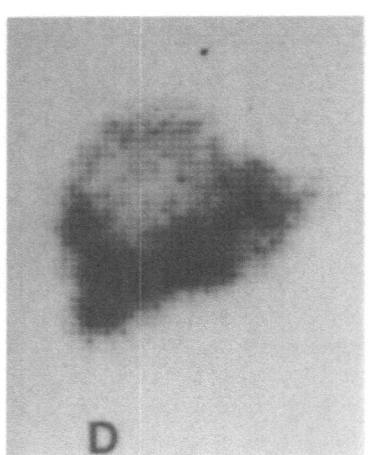

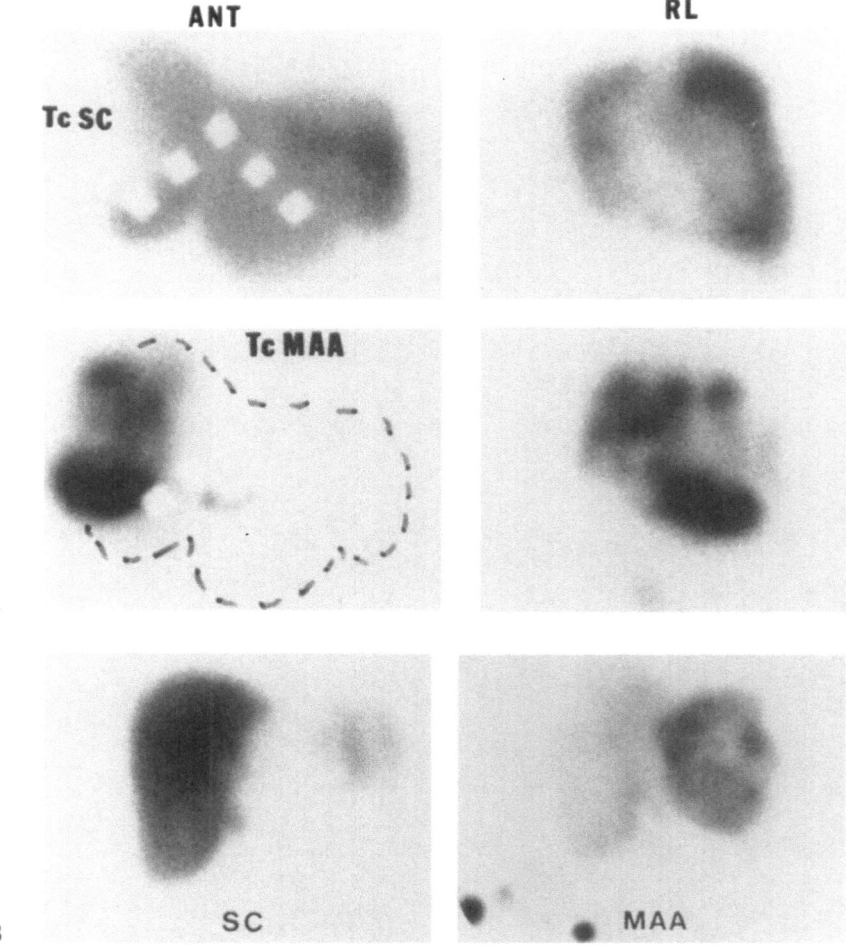

ANT

RL

Tc SC

Tc MAA

A

B   SC

MAA

Figure 4.8. High-tumor/nontumor ^{99m}**Tc-MAA perfusion ratio.**

There is considerably more perfusion to the tumor than to the remaining uninvolved liver.

A: Preoperative anterior (ANT) and right lateral (RL) ^{99m}Tc SC study (*above*) and postoperative ^{99m}Tc-MAA study (*below*) in the same patient as in Figure 4.5A. The ^{99m}Tc-MAA shows good perfusion of the tumor in the right lobe and minimal perfusion to the remainder of the liver.

Comment: This is ideal from a therapeutic advantage standpoint since most all the chemotherapy goes to the tumor with minimal exposure to normal liver.

(Reprinted from Thrall JH, Gyves JW, Ziessman HA, Ensminger WD. Hepatic arterial perfusion studies. In: *Diagnostic Interventions in Nuclear Medicine*. Chicago: Yearbook Medical Publishers; 1985:176–194. With permission.)

B: The ^{99m}Tc SC (SC) shows a large tumor defect in the left lobe of the liver. The ^{99m}Tc-MAA study (MAA) shows perfusion predominantly to the area of the tumor, although some minimal uptake is seen in the right lobe as well.

Figure 4.9. Hyperperfused tumor nodule with central necrosis.

A large tumor is seen in the midliver on the ^{99m}Tc SC study (*left*). The ^{99m}Tc-MAA study shows hyperperfusion predominantly to the rim of the tumor, particularly superolaterally, and a large cold area centrally.

Comment: As a tumor nodule enlarges, it outgrows the blood supply, resulting in central necrosis. Intraarterial chemotherapy can be effective here since the rapid growing portion of the tumor is the hyperperfused peripheral rim. This example also illustrates that proper positioning of the catheter does not always mean complete perfusion (e.g., nonperfusion of central necrosis). However, there may be a transition zone of hypoperfused but viable tumor cells. This has therapeutic implications since intraarterial therapy may eventually fail if chemotherapy cannot reach these tumor cells.

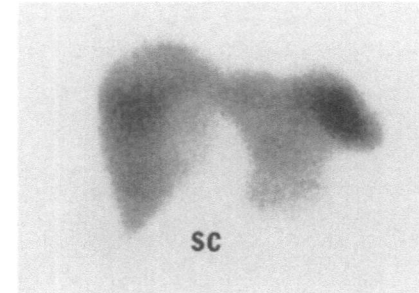 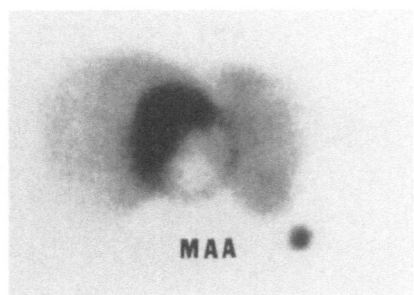

Figure 4.10. Correlative imaging in a patient with carcinoid tumor metastatic to the liver.

A ^{99m}Tc-MAA perfusion study (**A**, anterior view) shows nearly complete perfusion of both lobes of the liver with relative hyperperfusion of tumor nodules. These hyperperfused tumor nodules correspond to focal defects seen on the ^{99m}Tc SC scan (**B**). X-ray CT (**D**) shows a distribution of liver metastases similar to that seen on the ^{99m}Tc SC study. Contrast angiography (**C**) shows tumor nodules that are hypervascular peripherally and hypovascular centrally. This pattern is remarkably similar to that which is seen on the ^{99m}Tc-MAA perfusion study.

Comment: Carcinoid tumors are known to be vascular and routine contrast angiography will demonstrate this pattern. In contrast, colon cancer metastatic to the liver will usually appear "hypovascular" on routine angiography, but will demonstrate a similar pattern of hypervascularity if selective hepatic artery angiography is performed.[43]

(Reprinted from Ziessman HA, Gyves JW, Juni JE, et al. Atlas of hepatic arterial perfusion scintigraphy. *Clin Nucl Med*. 1985; 10:675–681. With permission.)

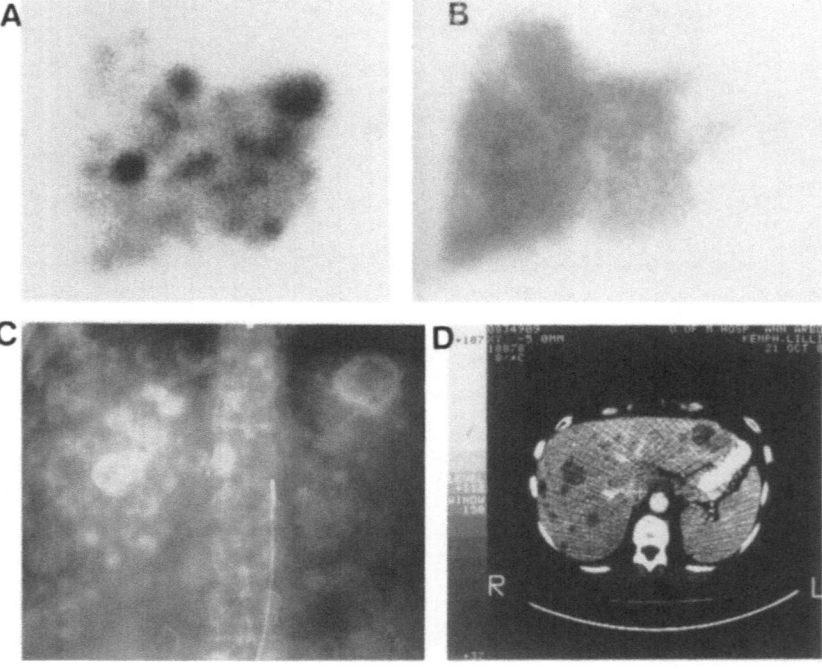

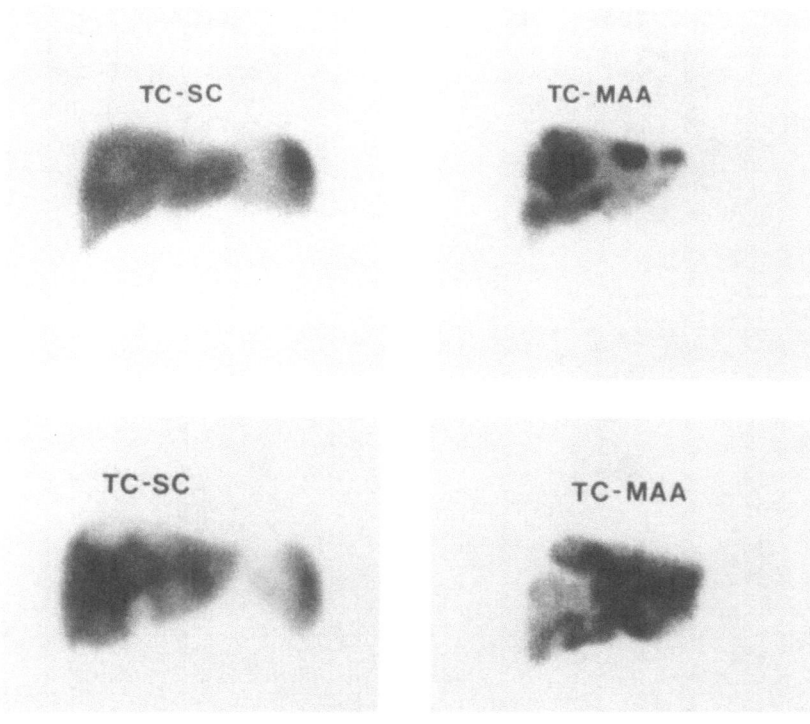

Figure 4.11. Complete perfusion of the liver: two patterns of tumor perfusion.

A: Hyperperfused solid tumor nodules on the ^{99m}Tc-MAA study correspond to relatively hypoperfused areas on ^{99m}Tc SC study.

B: Hyperperfused tumor nodules on ^{99m}Tc-MAA study correspond to relatively hypoperfused areas on Tc-SC study. However, note that many of these tumor nodules have a hyperperfused peripheral rim, but relatively hypoperfused central core.

Comment: Both patients (**A** and **B**) with metastatic colon cancer have complete perfusion of the entire tumor-bearing liver and both have relative hyperperfusion of tumor nodules. However, one has solid-appearing hyperperfused tumor nodules and the other has tumor nodules with a hyperperfused peripheral rim, but central hypoperfusion. Complete perfusion of the liver is obtainable in 88% of patients receiving surgically placed intraarterial catheters (93% in those with standard vascular anatomy and 79% of those with variant anatomy).[15] Experience in the placement of these catheters is very important.

(Reprinted from Ziessman HA, Thrall JH, Yang PJ, et al. Hepatic arterial perfusion scintigraphy with Tc-99m MAA. *Radiology*. 1984;152:167–172. With permission.)

Figure 4.12. Two catheters required to perfuse the entire tumor-bearing liver.

A: A HIDA study (*right*) performed 48 hr before the ^{99m}Tc-MAA study (*left*). The anterior view (ANT) is an early image (5 min after injection) before biliary drainage has occurred. The right lateral (RL) was taken at 60 min; therefore, the gallbladder is seen anteriorly and common duct/duodenum posteriorly. The tumor defects seen on the Tc-HIDA study in the left lobe generally correpond to the hyperperfused tumor nodules seen on ^{99m}Tc-MAA.

B: Two catheters are required to ensure complete perfusion of the entire tumor and liver. This is the same patient as in **A**. The two catheters were injected on separate days. One catheter is perfusing the right lobe and another perfusing predominantly the left lobe. An alternative method would be to use computer subtraction techniques so that both catheters could be injected on the same day (examples of computer subtraction in Figures 4.8B and 4.26A).

Comment: A ^{99m}Tc SC study is usually used for correlation with the ^{99m}Tc-MAA study, but since this patient had a recent Tc-HIDA study ordered for clinical reasons, it was used for correlation with the ^{99m}Tc-MAA study. The ^{99m}Tc SC study is preferred since it is easier to obtain multiple views and visualization of the spleen may also be helpful in determining the presence or absence of extrahepatic perfusion. This is illustrated in later examples.

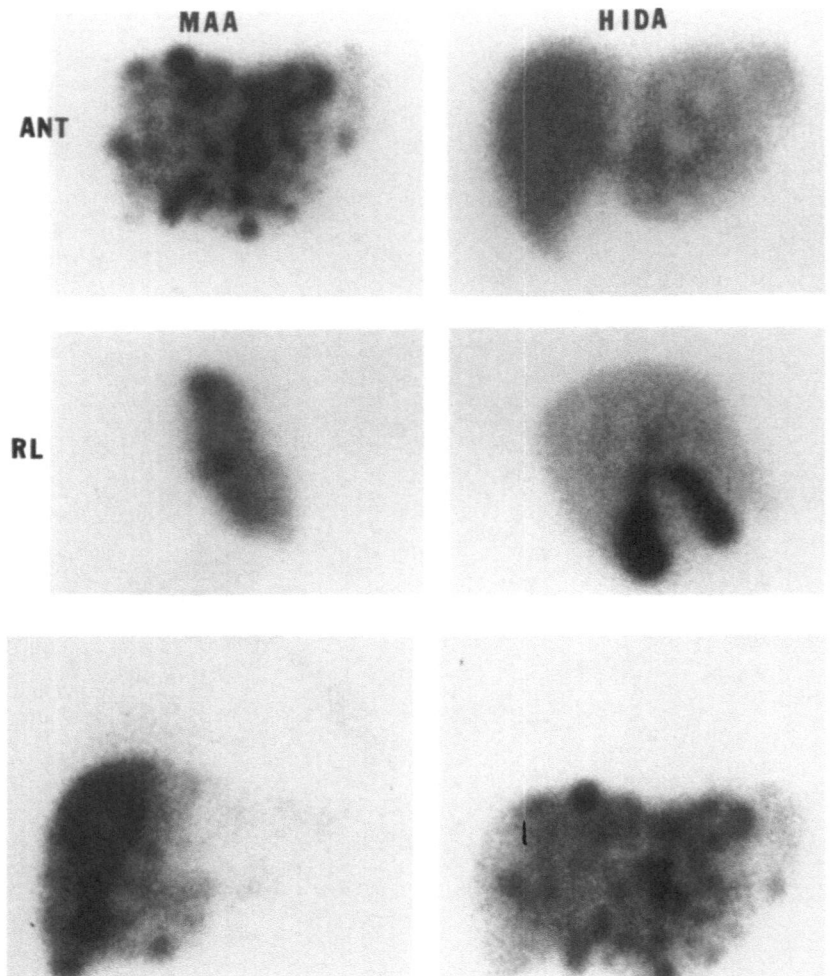

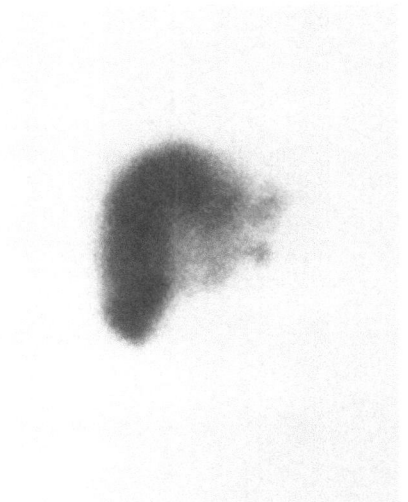 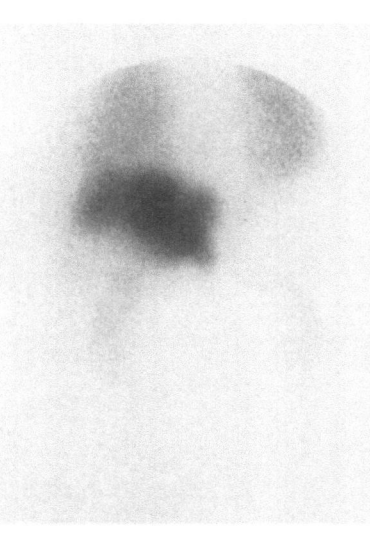

A

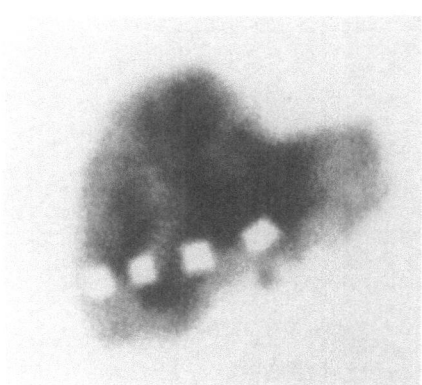 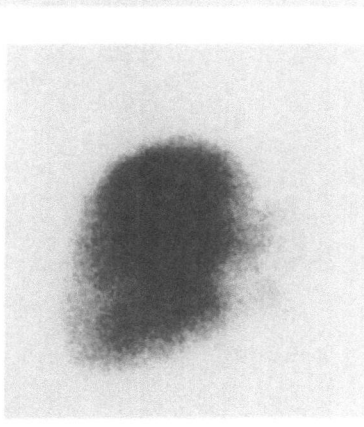

B

Figure 4.13. Change in the pattern of perfusion on follow-up scans.

A: The initial ^{99m}Tc-MAA image (*left*) shows essentially complete perfusion to both lobes of the liver after catheter placement. Two repeat studies over 6 months were unchanged. A follow-up scan at 13 months (*right*) demonstrates signficantly decreased perfusion to the right lobe as well as prominent A-V shunting to the lung. Faint renal visualization due to MAA breakdown or free pertechnetate is noted.

Comment: This change suggests catheter movement or partial catheter thrombosis. Catheter movement is not uncommmon with percutaneously placed catheters. With a fixed surgical catheter thrombosis is much more likely.

(Reprinted from Ziessman HA, Thrall JH, Yang PJ, et al. Hepatic arterial perfusion scintigraphy with Tc-99m MAA. *Radiology*. 1984;152:167–172. With permission.)

B: Initial study on another patient with metastatic colon cancer shows complete perfusion to both lobes of the liver (*left*). A follow-up scan (*right*) obtained 3 months later showed incomplete perfusion of the left lobe.

Comment: Intraarterial chemotherapy must be directed to the entire tumor-bearing liver to be effective. This patient's intraarterial therapy was continued, but concurrent intravenous chemotherapy was also started.

Figure 4.14. Arteriovenous (A-V) shunting to the lung.

A: There is good liver perfusion. However, a large amount of the ^{99m}Tc-MAA has been shunted through the liver to the lungs. The A-V shunt index (see **B**) was 32%.

Comment: This degree of shunting will result in less chemotherapy delivered to the tumor and increased systemic exposure to the drug, potentially leading to decreased antitumor effectiveness and adverse side effects.

(Reprinted from Ziessman HA, Gyves JW, Ensminger WD, et al. Quantitative hepatic arterial perfusion scintigraphy and starch microspheres in cancer chemotherapy. *J Nucl Med.* 1983;24:871–875. With permission.)

B: Quantitation of A-V shunting to the lungs. A region of interest is drawn by computer around the liver and lung in the anterior view. An A-V shunt index is calculated by dividing the lung region counts by the combined liver and lung counts. In this example (different patient), the shunt index = 3.2%.

Comment: Quantitation of lung uptake may be helpful clincially.[29,44] The mean percent shunt observed in 147 patients with surgically placed catheters was 6 ± 4% (range 0.4–32%).[15] Greater than 20% A-V shunting to the lung has been associated with an increased incidence of side effects.[44]

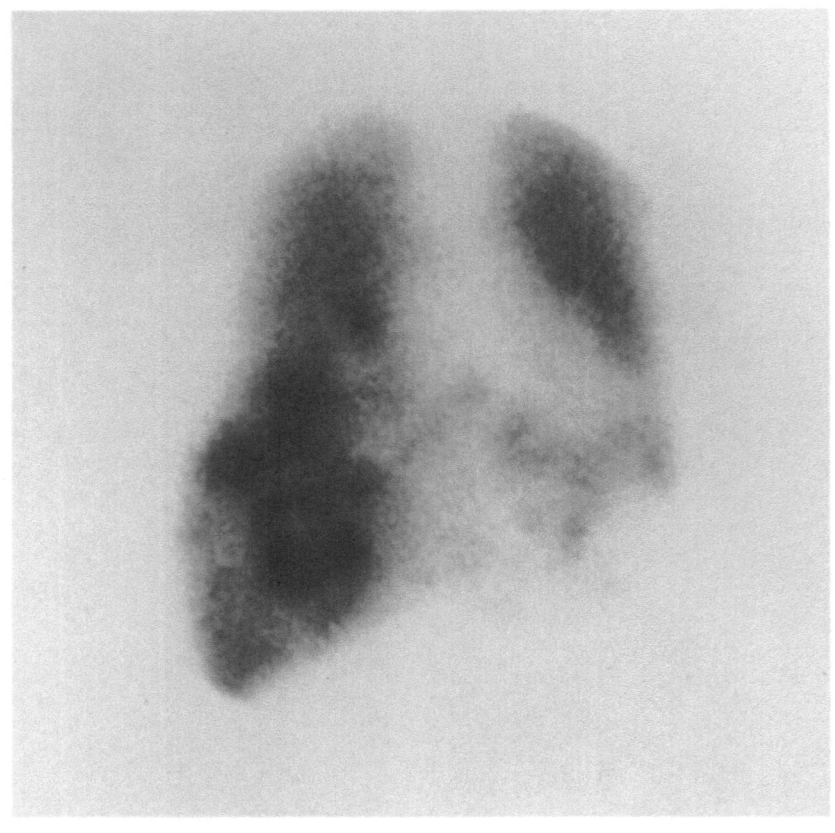

A

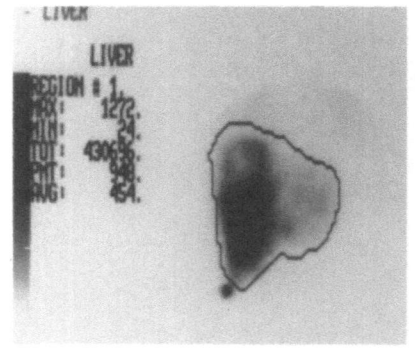
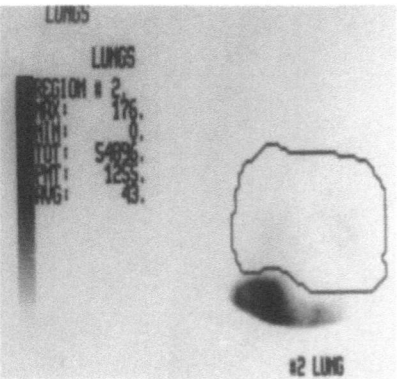

B

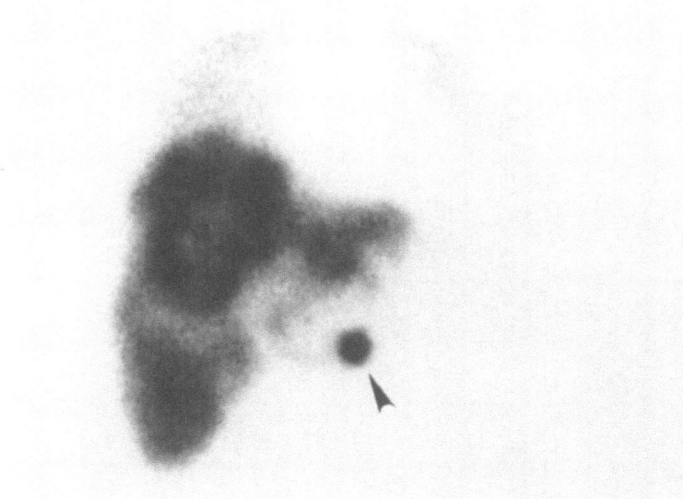

Figure 4.15. "Hot spot" at the catheter tip due to adherance of ^{99m}Tc-MAA to a non-obstructive clot similar to that seen on ^{99m}Tc-MAA venograms with venous thrombosis.

There is good perfusion to the liver, a small amount of shunting to the lung, and a hot spot in the region of the catheter tip (*arrowhead*).

Comment: A hot spot at the catheter tip is seen in approximately 10% of patients after surgical placement.[15] It is due to local clot formation and usually resolves spontaneously. By itself, it is not associated with adverse symptoms or course. However, decreased liver perfusion and/or evidence of extrahepatic perfusion associated with a hot spot is strongly suggestive of catheter thrombosis. New symptoms suggestive of gastrointestinal toxicity would be an indication to repeat this study to see if progressive thrombosis has occurred.

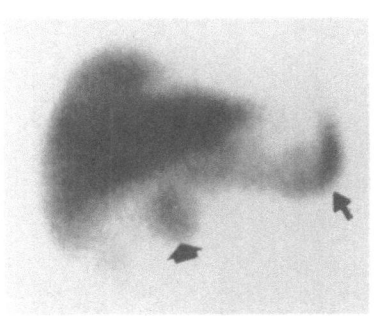 

Figure 4.16. Extrahepatic gastrointestinal perfusion due to incorrect positioning of a percutaneously placed catheter.

Although the immediate postcatheter placement ^{99m}Tc-MAA study (*left*) shows good perfusion to both lobes of the liver, there is obvious perfusion to the stomach (*arrow*) and duodenum (*arrowhead*). The patient's catheter was repositioned (*right*) and a follow-up ^{99m}Tc-MAA showed no extrahepatic perfusion. Liver perfusion remained good.

Comment: Extrahepatic perfusion is seen more frequently with percutaneously placed catheters (51%) compared with surgically placed catheters (14%).[15,19,45]

(Reprinted from Ziessman HA, Gyves JW, Juni JE, et al. Atlas of hepatic arterial perfusion scintigraphy. *Clin Nucl Med.* 1985; 10:675–681. With permission.)

Figure 4.17. Good liver perfusion with definite extrahepatic gastroinestinal perfusion.

There is certain perfusion of the stomach (*open arrowhead*), duodenum (*closed arrowhead*) and, to a lesser extent, bowel (*open arrows*). The liver is well perfused. Focal uptake below the right lobe is seen within the infusion pump port.

Comment: Extrahepatic perfusion is frequently associated with side effects. In one study, 45% of patients with gastrointestinal uptake experienced epigastric pain or gastritis compared with only 16% of those without gastrointestinal uptake.[16] Another study in patients with surgically placed catheters found that 70% of patients with extrahepatic perfusion on ^{99m}Tc-MAA studies had significant clinical symptoms (abdominal pain, gastritis, ulcers, gastrointestinal bleeding) that required adjustment of dosage, surgical repositioning of the catheter, or discontinuation of intraarterial chemotherapy. Patients with no evidence of extrahepatic perfusion had only a 19% incidence of similar symptoms.[15]

(Reprinted from Ziessman HA, Thrall JH, Yang PJ, et al. Hepatic arterial perfusion scintigraphy with Tc-99m MAA. *Radiology*. 1984;152:167–172. With permission.)

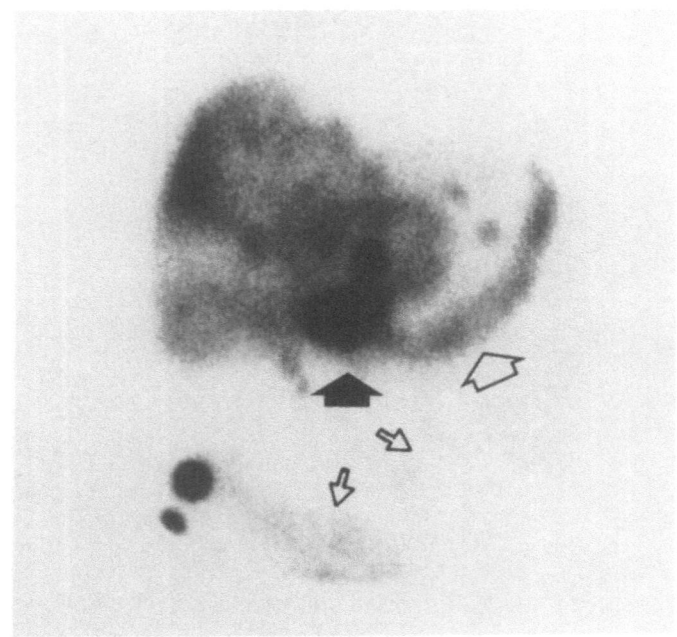

Figure 4.18. Definite extrahepatic perfusion and decreased liver perfusion.

A: Prominent extrahepatic perfusion to the stomach, small bowel, and spleen on the ^{99m}Tc-MAA study (*right*). The liver has less perfusion than the other visceral organs. In this case the tumor nodules do not seem hyperperfused. In fact, they appear hypoperfused. However, careful comparison with the ^{99m}Tc SC study reveals that the photopenic area on the ^{99m}Tc-MAA study is smaller than on the ^{99m}Tc SC study. This is because there is a peripheral rim of perfused tumor.

Comment: Sometimes extrahepatic perfusion is obvious on perfusion imaging. It is most commonly seen in the stomach and spleen; however, inadvertent perfusion to the gallbladder, pancreas, and bowel may occur as well.

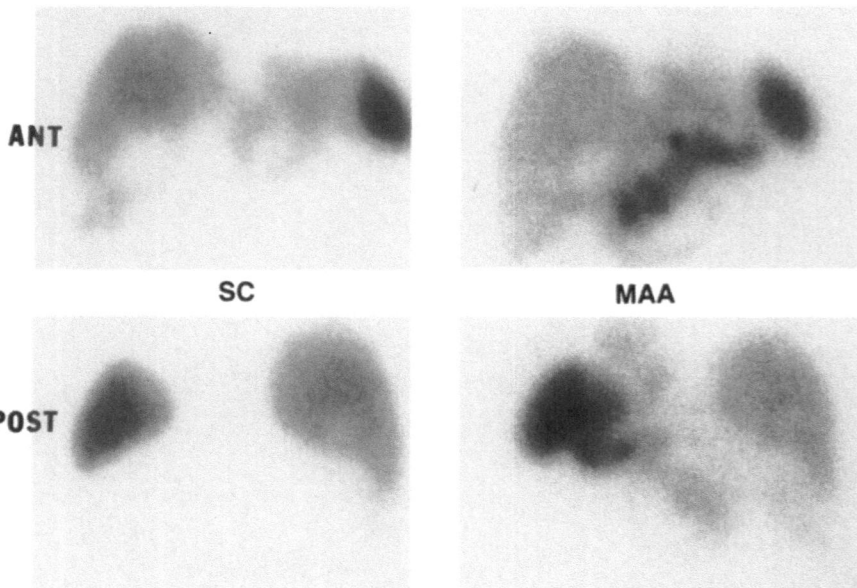

A

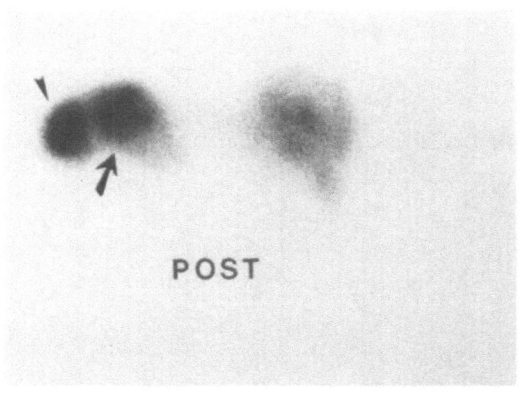

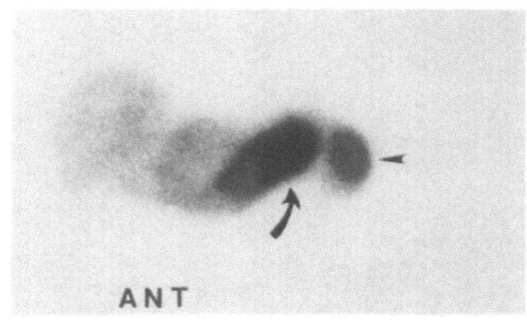

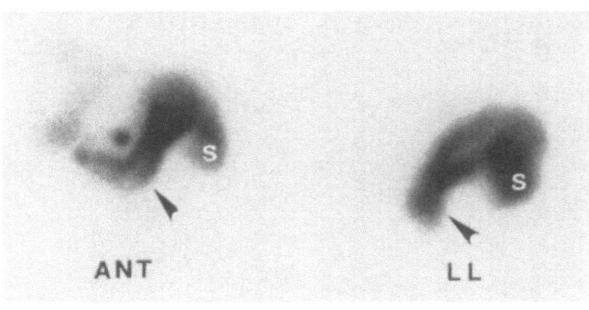

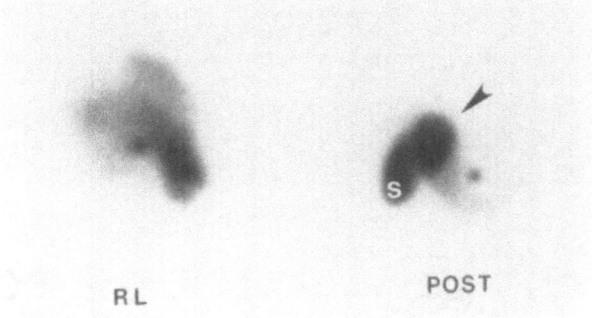

This case is unusual in that the amount of tumor perfusion appears to be similar to surrounding liver. This may be more apparent than real. Overlying liver activity may mask relative hyperperfusion of the peripheral portion of the tumor. SPECT can improve contrast resolution and usually shows a hyperperfused rim of tumor even when the planar images do not (see Figures 4.27, 4.28).[45]

B: Very decreased perfusion to the liver and definite extrahepatic perfusion of the stomach (*arrow*) and spleen (*arrowhead*).

Comment: The surgeon who had placed this catheter was inexperienced in this technique. The ^{99m}Tc-MAA study was performed 24 hr after surgery.

C: Predominant extrahepatic perfusion. There is a hot spot in the region of the catheter tip in addition to very decreased liver perfusion and extensive extrahepatic perfusion to the stomach (*arrowheads*) and spleen (S).

Comment: Multiple views can increase one's certainty about the degree of extrahepatic perfusion and the exact organs involved. This combination of findings almost always represents hepatic arterial thrombosis with retrograde or collateral flow.

(Reprinted from Ziessman HA, Thrall JH, Yang PJ, et al. Hepatic arterial pefusion scintigraphy with Tc-99m MAA. *Radiology*. 1984;152:167–172. With permission.)

Figure 4.19. The use of multiple views and correlation with ^{99m}Tc SC for less certain gastric perfusion.

A: The activity adjacent to the left lobe has the appearance of gastric perfusion on the anterior view (*arrowhead*). However, without the correlative ^{99m}Tc SC study one cannot be sure this is not tumor hyperperfusion. Comparison of the two anterior images (^{99m}Tc SC and ^{99m}Tc-MAA) shows that the suspicious activity is clearly adjacent to the left lobe where the stomach lies. The left lateral view shows activity just behind the left lobe (*arrow*) and the posterior view shows the curvilinear activity in a pattern also suggestive of stomach (*arrow*). Note the hot spot in the region of the catheter tip suggestive of thrombosis.

B: Suggestive of gastric perfusion on the anterior view (*arrrowhead*). Note the adjacent hyperperfused tumor nodule. Again the left lateral (LL) and posterior (POST) views help confirm that this is stomach. On the left lateral view there is gastric perfusion (*arrowhead*) immediately behind the left lobe and anterior to the spleen (S). The posterior view also shows characteristic gastric perfuson (*arrowhead*).

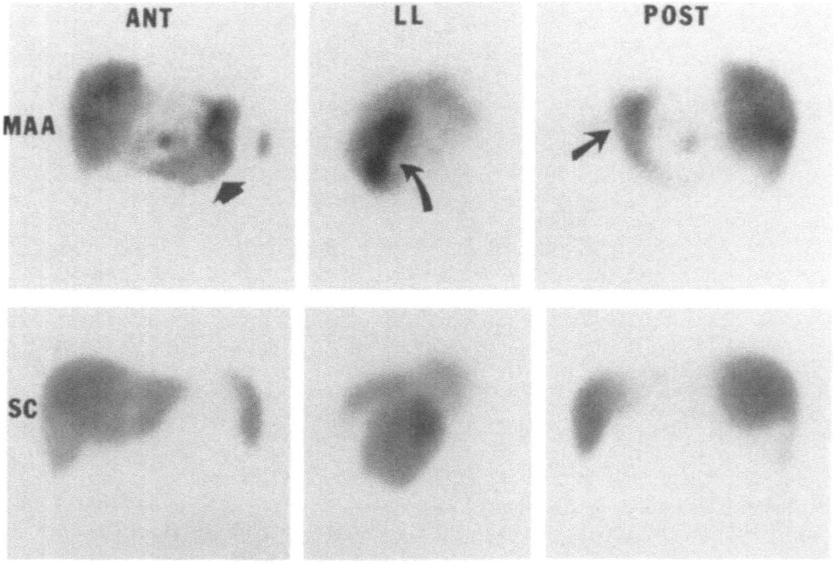

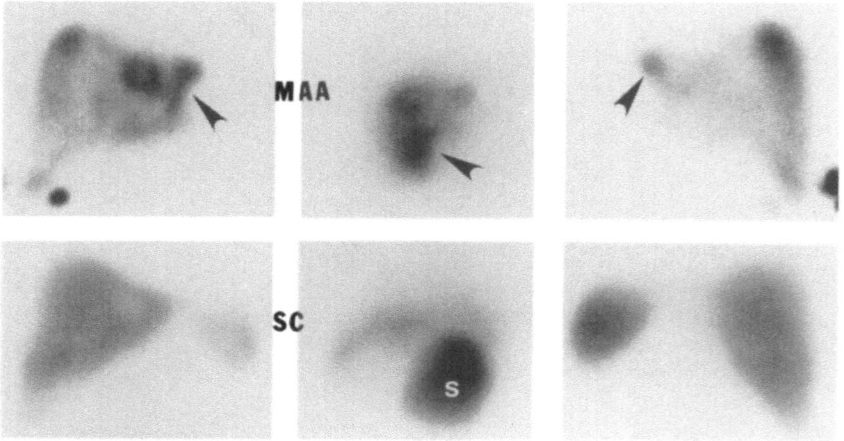

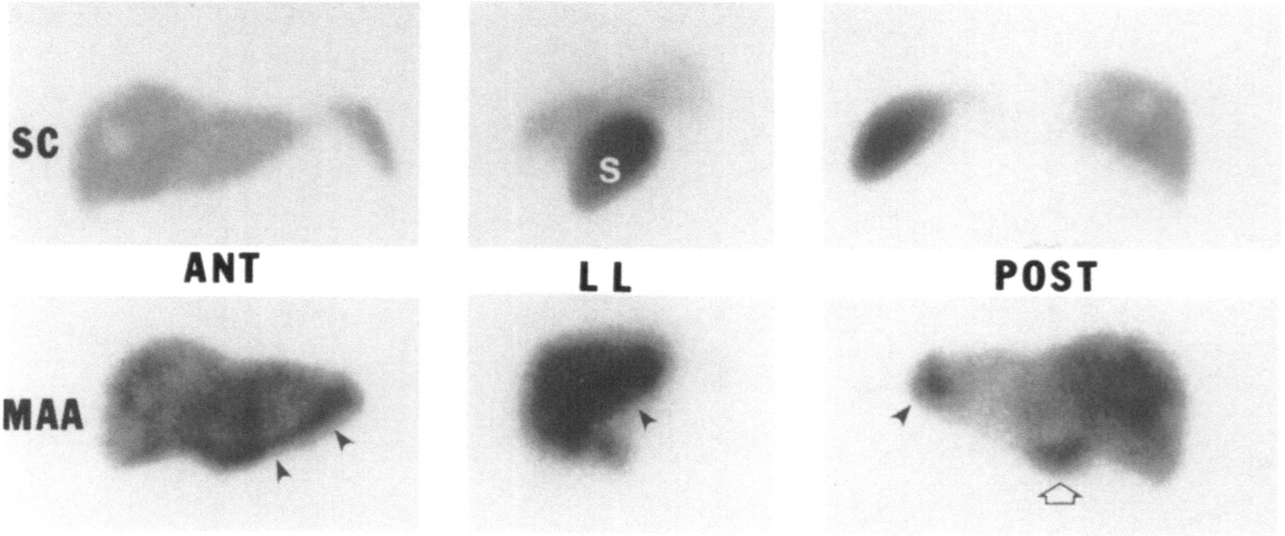

Comment: Extrahepatic perfusion resulting in adverse symptomatology may not always be immediately apparent on ^{99m}Tc-MAA imaging. A recent ^{99m}Tc SC study is exceedingly helpful. Extrahepatic perfusion that is easily noted when the two studies are compared may be missed entirely if a ^{99m}Tc SC scan is not available at the time of inter-

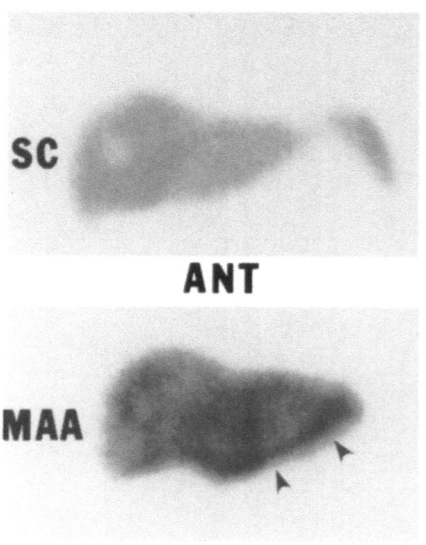

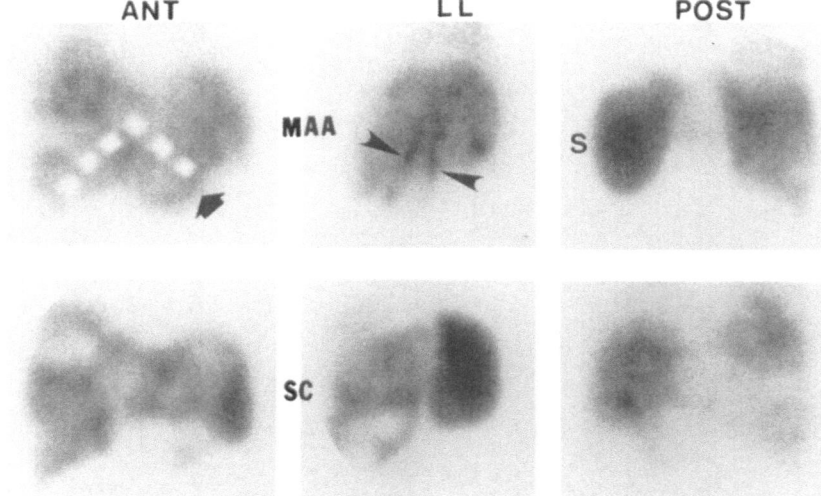

pretation. Anatomical boundaries may be poorly defined on the ^{99m}Tc-MAA study alone due to overlapping activity (e.g., the left lobe and the stomach). Multiple views, including the anterior, posterior, right lateral, and left lateral are particularly helpful in defining the extent of hepatic and extrahepatic perfusion.[15]

C: Very subtle gastric perfusion. Note the curvilinear gastric activity (*arrowheads*) at the inferior medial aspect of the left lobe on the anterior view, behind the left lobe, and anterior to the spleen (S) on the left lateral view. The posterior view also shows activity in the region of the stomach (*arrowhead*) and duodenum (*open arrowhead*).

Comment: Multiple views are mandatory here. Even subtle extrahepatic perfusion as seen here on the ^{99m}Tc-MAA perfusion study may be associated with clinically significant side effects.

(Reprinted from Ziessman HA, Gyves JW, Juni JE, et al. Atlas of hepatic arterial perfusion scintigraphy. *Clin Nucl Med.* 1985; 10:675–681. With permission.)

D: Extremely subtle gastric perfusion associated with symptoms of nausea and vomiting. The anterior view shows activity that could easily be due to tumor perfusion, although there is a slight curvilinear stomach-like appearance to it (*arrowhead*). The posterior view shows definite splenic (S) uptake.

Comment: Whenever splenic perfusion is seen, gastric perfusion can also be found if looked for carefully. On the left lateral view, linear activity can be seen along the anterior and posterior walls of the stomach (*arrows*).

(Reprinted from Ziessman HA, Gyves JW, Juni JE, et al. Atlas of hepatic arterial perfusion scintigraphy. *Clin Nucl Med.* 1985; 10:675–681. With permission.)

Figure 4.20. Changing pattern of extrahepatic gastric perfusion.

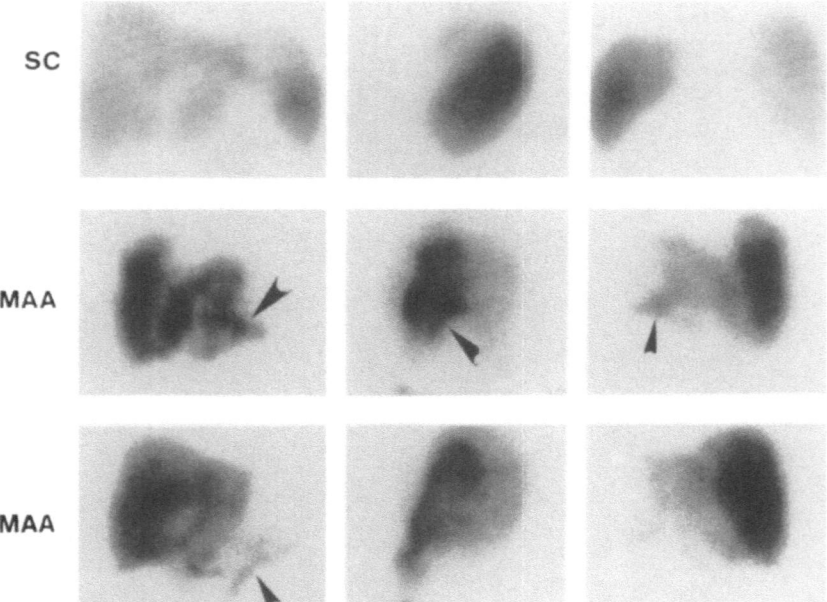

Top: 99mTc SC (SC) study. Tumor involvement is not obvious here, but it was certain on CT (not shown).

Middle: Postcatheter placement. Irregular linear activity in the left lobe on the anterior view and activity behind the left lobe on the left lateral (*arrowheads*) suspicious but not diagnostic of gastric perfusion. The posterior view also has uptake in the region of the distal stomach (*arrow*). It was decided to start therapy and observe for adverse effects. Because of persistent nausea and vomiting the patient had a follow-up study (*bottom*).

Bottom: The follow-up 99mTc-MAA study shows definite extrahepatic perfusion to the stomach (*arrowhead*) confirming suspicions on the prior study. Liver perfusion seems improved on this study. The infusion rate of chemotherapy was decreased with improvement of symptoms.

Comment: It may be difficult clinically to determine whether a patient's adverse symptoms are due to tumor progression and nonresponse to chemotherapy or, alternatively, due to poor positioning of the catheter so the therapy is not optimally delivered to the tumor, but rather to extrahepatic abdominal organs. 99mTc-MAA studies are extremely useful in making this important differential diagnosis.[14-16] It is important to remember that the pattern of perfusion may change during the course of therapy due to catheter movement or thrombosis.

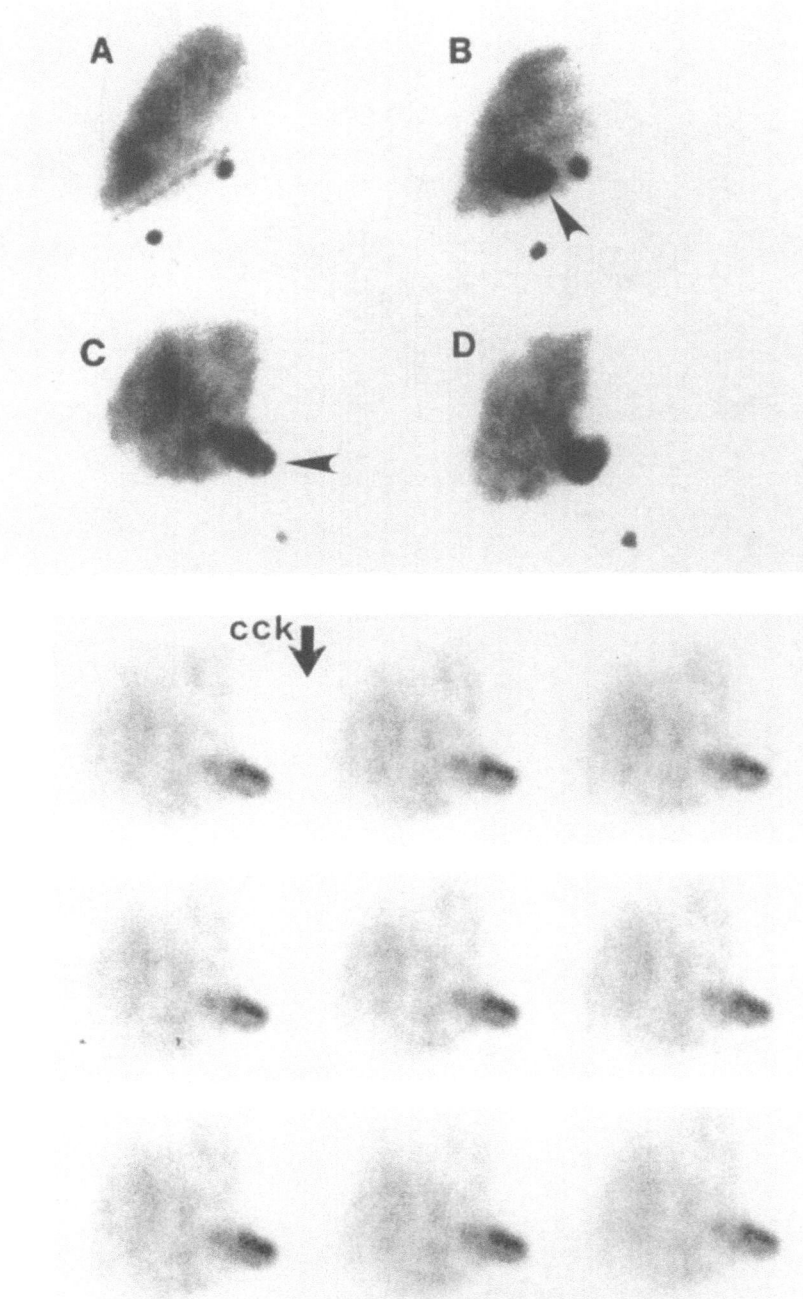

A

B

Figure 4.21. Gallbladder perfusion.

A: The patient's initial ^{99m}Tc-MAA study (A) shows perfusion limited to the right lobe of the liver. The focal hot spot near the xiphoid represents a clot in the catheter tip while the other hot spot below is within the sideport of the infusion pump. A repeat study in the anterior (B), right lateral (C), and right anterior oblique (D) views was performed 4 months later for symptoms of nausea and right subscapular pain starting with the most recent course of intraarterial chemotherapy. This study shows prominent perfusion of the gallbladder (*arrowhead*).
(Reprinted from Ziessman HA, Gyves JW, Juni JE, et al. Atlas of hepatic arterial perfusion scintigraphy. *Clin Nucl Med.* 1985; 10:675–681. With permission.)
B: Gallbladder contraction with cholecystokinin. Images every 2 min for 18 min shown after intravenous infusion of cholecystokinin (Kinevac, E.R. Squibb & Sons, Princeton, NJ). Note progressive contraction of the gallbladder wall, maximal on the last image (*arrow*). This shows subnormal contraction of the inflamed hyperperfused gallbladder wall.
Comment: The cystic artery normally originates from the right hepatic artery. As a result, chemical cholecystitis frequently occurs in patients receiving hepatic arterial chemotherapy. Elective cholecystectomy is now performed routinely at the time of surgical catheter placement.
(Reprinted from Ziessman HA, Gyves JW, Juni JE, et al. Atlas of hepatic arterial perfusion scintigraphy. *Clin Nucl Med.* 1985; 10:675–681. With permission.)

Figure 4.22. Pancreatic scan.

This ^{99m}Tc-MAA study shows perfusion to
the liver but also prominent extrahepatic
perfusion to the pancreas. There is some
perfusion to the stomach (*open arrowhead*,
hard to see on this image) and spleen
(*closed arrowhead*) as well.

Comment: This is an unacceptable perfusion pattern and intraarterial chemotherapy
had to be discontinued.

(Reprinted from Ziessman HA, Gyves JW,
Juni JE, et al. Atlas of hepatic arterial perfusion scintigraphy. *Clin Nucl Med*. 1985;
10:675–681. With permission.)

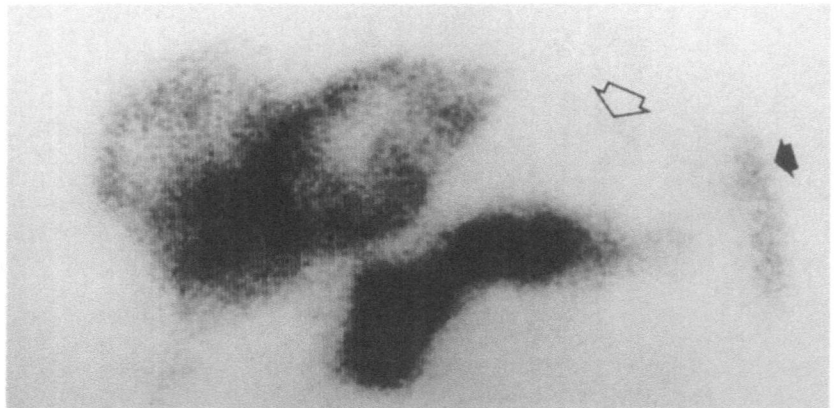

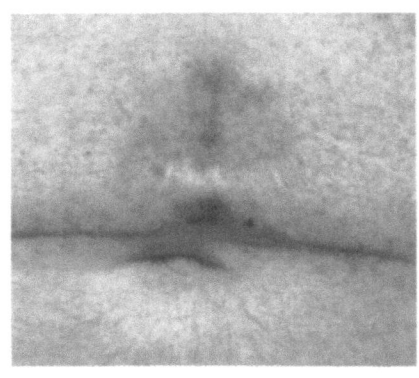 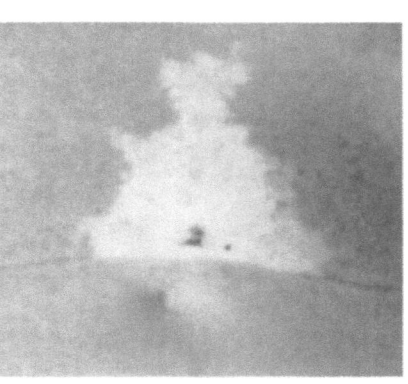

A

ANT   LL

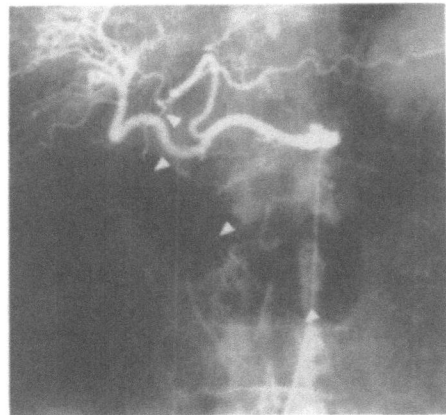 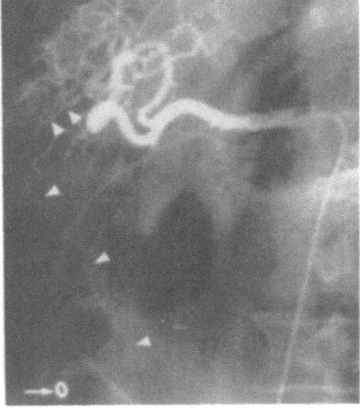

SC

MAA

MAA

B

C

Figure 4.23. Supraumbilical skin rash after each course of chemotherapy due to perfusion of the anterior abdominal wall by the falciform artery.

A: *Left:* Photograph of midline supraumbilical rash. *Right:* Photograph with ultraviolet light after injection of 5 ml fluorescein sodium in the sideport of the infusion pump. The region of fluorescence coincides with the cutaneous rash.

B: *Above:* ^{99m}Tc SC study (ANT and LL) for comparison. *Middle:* ^{99m}Tc-MAA study 3 weeks after catheter and pump placement, after patient noted a supraumbilical rash. Anterior and left lateral views showed marked extrahepatic perfusion in the distribution of falciform ligament (*arrowheads*). The location of the anterior extension of activity just beaneath the supraumbilical rash was confirmed by a radionuclide skin marker (*arrow*).

C: Postoperative celiac arteriogram in anteroposterior (*left*) and right posterior oblique (*right*) projections. Lead skin marker (arrow-O) marks supraumbilical rash. The splenic artery has been ligated. Stump of ligated gastroduodenal artery is visible just proximal to left hepatic artery. The falciform artery (*arrowheads*) arises from the left hepatic artery and follows a characteristic L-shaped caudal course anterolateral to the anterior surface of the liver, then caudal and medial to the posterior surface of the anterior abdominal wall at the L3–4 level, ending in a network of fine vessels just beneath the marker. A small filling defect, probably a clot, is present at the infusion catheter tip.

(Reprinted from Williams DM, Cho KJ, Ensminger WD, Ziessman HA, Gyves JW. Hepatic falciform artery: anatomy, angiographic appearance, and clinical significance. *Radiology.* 1985;156:339–340. With permission.)

Figure 4.24A–G. Gastric air contrast ^{99m}Tc-MAA study to aid in the diagnosis of gastric perfusion.

Despite multiple ^{99m}Tc-MAA views and a recent ^{99m}Tc SC liver–spleen scan for comparison, the presence or absence of gastric perfusion may sometimes be uncertain. Sodium bicarbonate–simethicone effervescent granules (E-Z Gas, Sparkles) can be quite helpful in making this differential. When ingested orally with water, 450 cc of carbon dioxide gas is generated, distending the stomach and producing a characteristic change in the pattern of perfusion.[39] We include this as part of our routine study.

A: Obvious gastric and duodenal perfusion used as an example of how the effervescent granules work. After ingestion of the effervescent granules, the perfused greater curvature of the stomach moves inferiorly and laterally, confirming gastric perfusion.

B: Gastric perfusion is seen only with the effervescent granules. The initial study (*left*) is not particularly suggestive of stomach perfusion. After ingestion, clear gastric perfusion (*arrow*) is seen.

(Reprinted from Ziessman HA, Gyves JW, Juni JE, et al. Atlas of hepatic arterial perfusion scintigraphy. *Clin Nucl Med.* 1985; 10:675–681. With permission.)

C: *Left column:* Anterior (*above*) and left lateral (*below*) views of ^{99m}Tc-MAA study raises the question of perfusion to the stomach in left upper quadrant versus a large left lobe. After administration of E-Z Gas, there is distention of fundus of the stomach (*arrows*) with clear delineation of stomach walls on the left lateral (*arrowheads*). To validate the effervescent granule study, Tc-DTPA was ingested orally. Increased activity is then seen within the stomach corresponding to changes seen on E-Z Gas study confirming gastric perfusion.

(Reprinted from Wahl RL, Ziessman HA, Juni J, Lahti D. Gastric air contrast: useful adjunct to hepatic artery scintigraphy. *AJR.* 1984;143:321–325. With permission.)

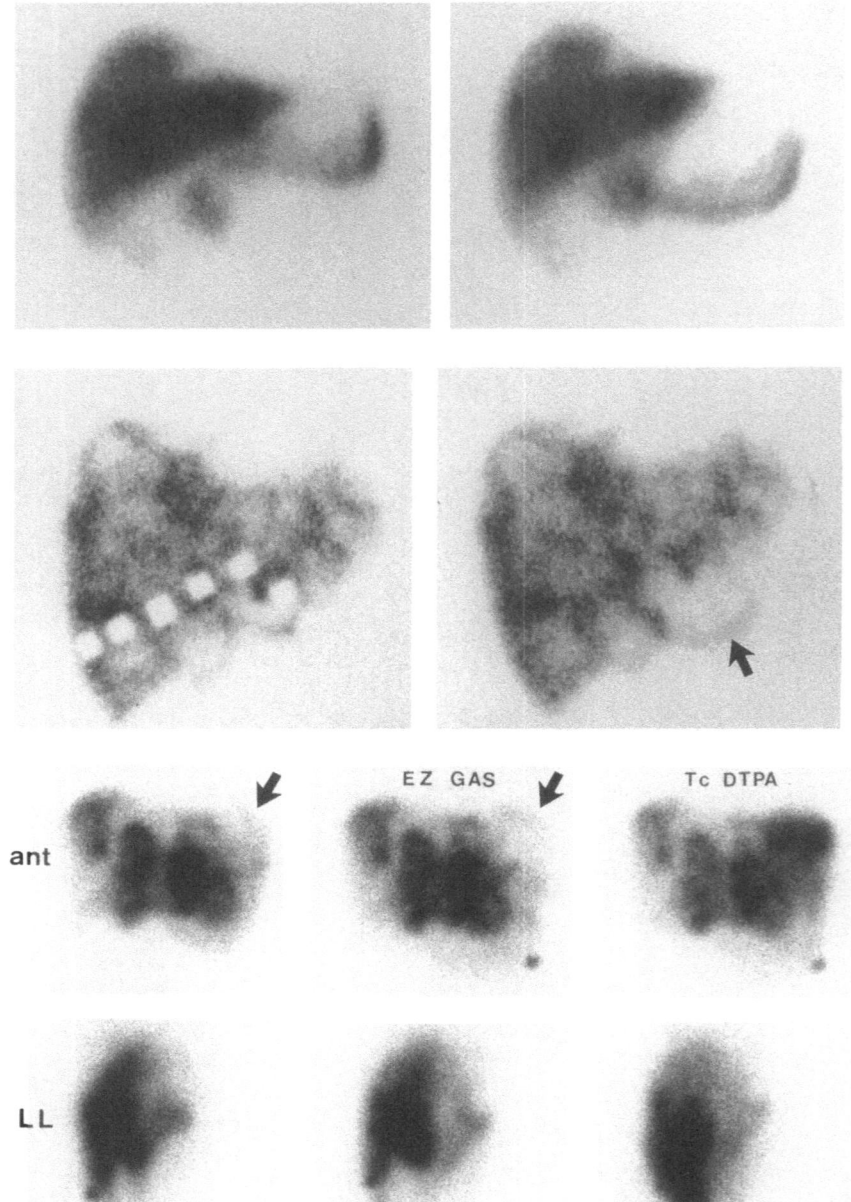

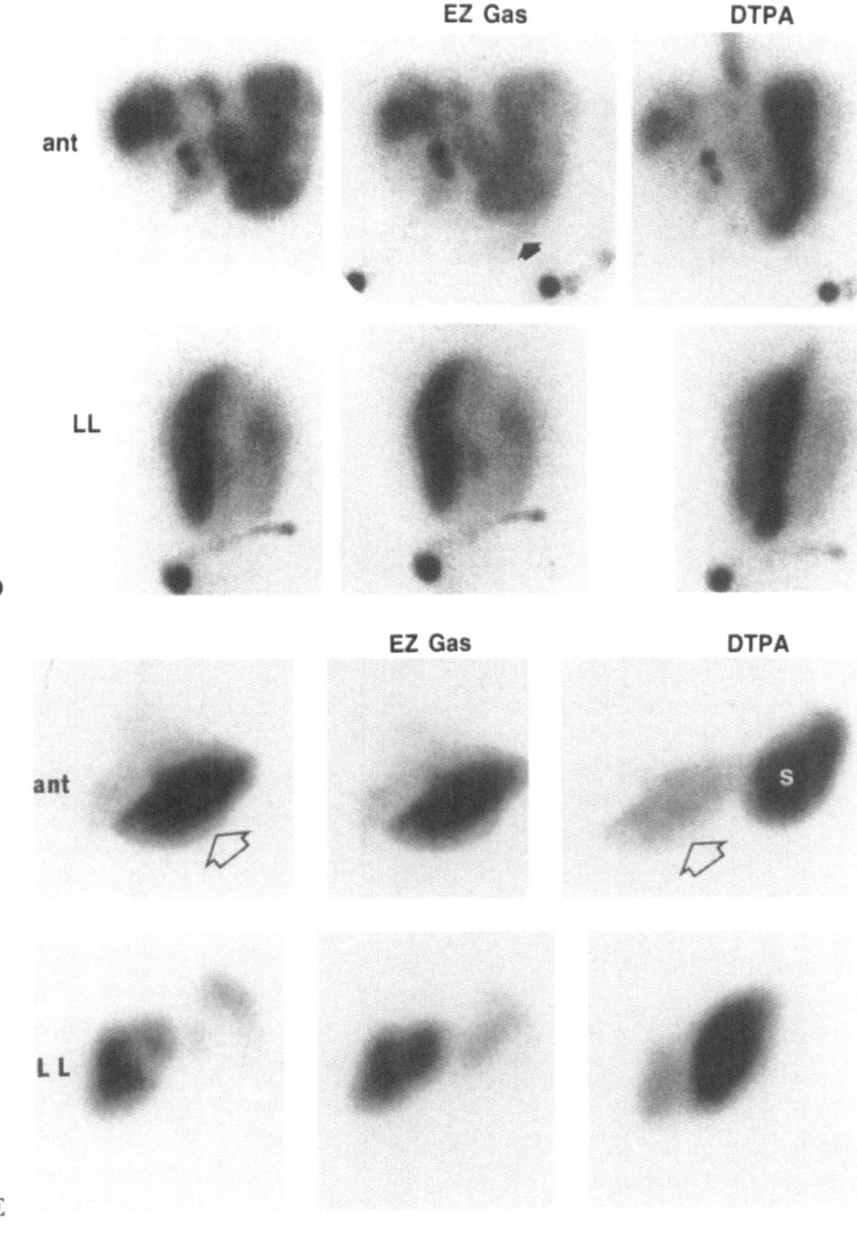

D: Anterior (ant) and left lateral (LL) views on ^{99m}Tc-MAA study raises the question of perfusion to stomach versus a large left lobe. *Middle column:* After 4 g of E-Z Gas crystals and water, similar views were obtained that show a subtle change in the left upper quadrant. A short portion of perfused gastric mucosa, probably a portion of the greater curvature of the stomach (*arrowhead*), is now seen indicating some gastric perfusion. Tc-DTPA, 1 mCi (37 MBq) in 100 ml water orally was confirmatory. This patient had mild syptoms of nausea that was relieved by lowering the dose of chemotherapy.

(Reprinted from Wahl RL, Ziessman HA, Juni J, Lahti D. Gastric air contrast: useful adjunct to hepatic artery scintigraphy. *AJR*. 1984;143:321–325. With permission.)

E: Negative for gastric perfusion. Anterior (ant) and left lateral (LL) views from ^{99m}Tc-MAA study show increased perfusion in the region of the left lobe (*open arrowhead*) of liver versus stomach, with lesser perfusion of the right lobe. After administration of E-Z Gas no change is seen in the appearance of activity, implying that this is not stomach and no extrahepatic perfusion is present. Oral Tc-DTPA confirms that the stomach (S) is medial to the left lobe (*arrowhead*). Thus, no gastric extrahepatic perfusion was present.

(Reprinted from Wahl RL, Ziessman HA, Juni J, Lahti D. Gastric air contrast: useful adjunct to hepatic artery scintigraphy. *AJR*. 1984;143:321–325. With permission.)

(Figure 4.24 continued on next page)

F: Water ^{99m}Tc-MAA perfusion study. A change in the initial pattern of perfusion in the left upper quadrant is seen after the ingestion of a glass of water. This patient refused to swallow the sodium bicarbonate effervescent granules because she experienced nausea (an occasional side effect of E-Z Gas) on a previous study. In this case, gastric perfusion is strongly suspected in the initial anterior image (*arrow*) and is confirmed by distention of the stomach after the water (*arrows*).

Comment: In some cases, the left lateral projection is the best for visualizing gastric perfusion; in other cases the anterior projection is best. A carbonated drink could also be used.

(Reprinted from Ziessman HA, Gyves JW, Juni JE, et al. Atlas of hepatic arterial perfusion scintigraphy. *Clin Nucl Med.* 1985; 10:675–681. With permission.)

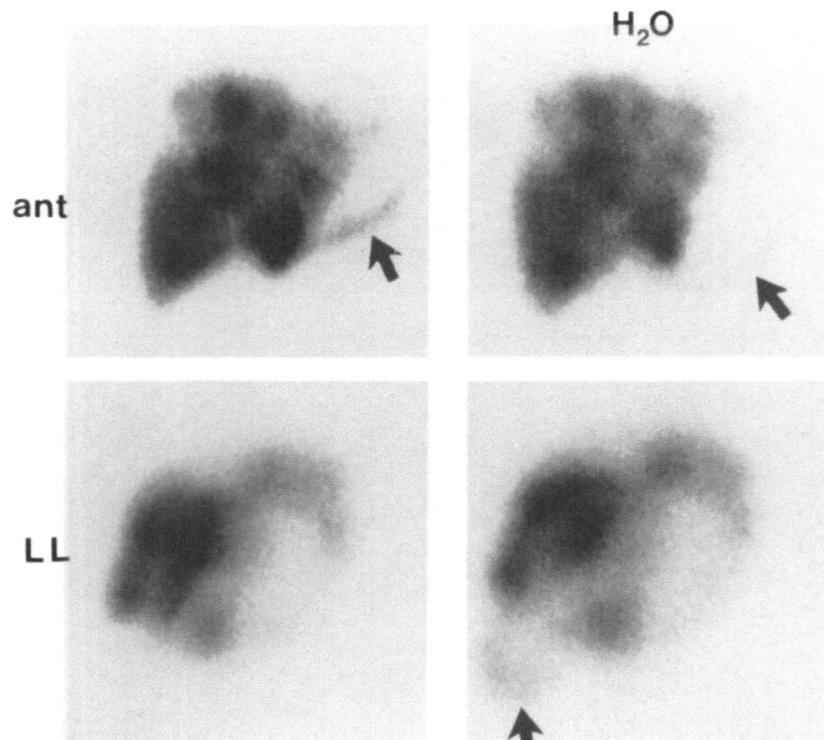

SC   MAA   EZ Gas

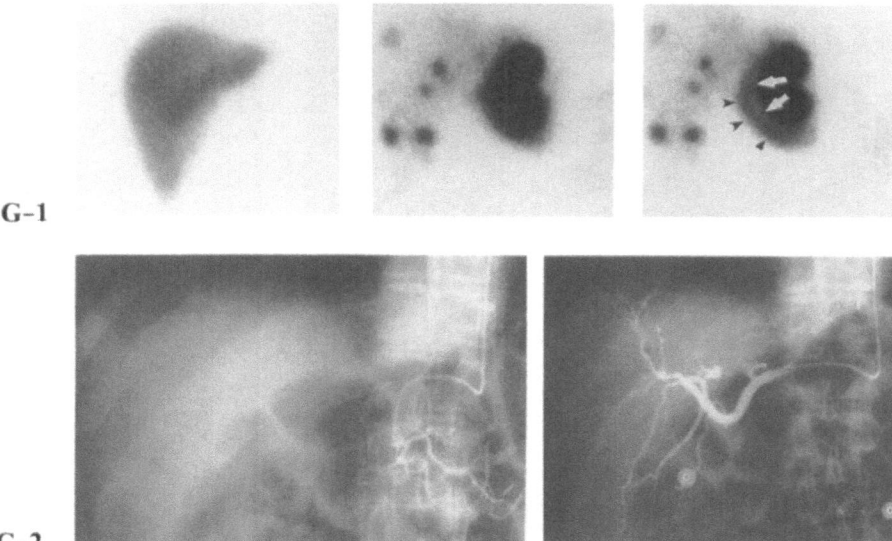

G–1

G–2

G–1: *Left:* ^{99m}Tc SC scan of patient who had undergone gastrectomy, splenectomy, and partial pancreatectomy for gastric carcinoma. *Middle:* In addition to hypervascular nodules in the right lobe, the ^{99m}Tc-MAA study shows markedly increased activity in the region of the left lobe, which could represent a large hepatic metastasis or bowel perfusion. *Right:* After E-Z Gas, a clear space (*white arrows*) is seen medial to the increased activity (*black arrowheads*), consistent with distension of small bowel by the effervescent granules. This was confirmed at angiography (see **G–2**).
(Reprinted from Wahl RL, Ziessman HA, Juni J, Lahti D. Gastric air contrast: useful adjunct to hepatic artery scintigraphy. *AJR.* 1984;143:321–325. With permission.)
G–2: *Left:* Angiogram of patient in **G–1** shows catheter malpositioned in gastroduodenal artery, corresponding spatially to perfused loop of duodenum in **G–1** (*black arrowheads*). *Right:* Catheter has been repositioned in hepatic artery distal to take-off of gastroduodenal artery, with resulting flow into the hepatic arteries.
(Reprinted from Wahl RL, Ziessman HA, Juni J, Lahti D. Gastric air contrast: useful adjunct to hepatic artery scintigraphy. *AJR.* 1984;143:321–325. With permission.)

Figure 4.25. Streptokinase infusion for hepatic arterial thrombosis.

A: Forty-eight hours after surgical placement of a hepatic arterial catheter, this patient developed right upper quadrant pain, fever, and a rapid rise in liver enzymes. A ^{99m}Tc-MAA study (*upper left*) revealed almost no perfusion to the liver, a hot spot in the region of the catheter tip suggestive of thrombosis, and definite extrahepatic perfusion to the stomach (*arrowhead*) and spleen (S). Contrast angiography via the femoral artery showed occlusion of the common hepatic and proper hepatic arteries. Streptokinase was then locally infused through the femoral artery catheter. Repeat angiography 48 hr later revealed complete recanalization of the arteries. Follow-up perfusion studies demonstrate unobstructed hepatic arterial flow with reperfusion to most of the liver. The perfusion pattern continued to improve over time on follow-up scans. By 9/19 the liver is well perfused and there is no gastric or splenic perfusion. (Reprinted from Ziessman HA, Gyves JW, Juni JE, et al. Atlas of hepatic arterial perfusion scintigraphy. *Clin Nucl Med.* 1985; 10:675–681. With permission.)

B-1: Patient with adenocarcinoma of rectum metastatic to liver developed right upper quadrant pain, fever, and rise in enzymes 3 days after catheter placement. Selective hepatic angiography demonstrated occlusion of the proper hepatic artery with collateral filling. Streptokinase was then infused via a left brachial angiographic catheter directly to the area of thrombosis. Within 24 hr, the proper hepatic was partially opened and by 48 hr, the right hepatic was widely patent without thrombosis, but the left remained occluded, although reconstructed by collaterals. Anterior, right lateral, and posterior views are shown.

Top: Three-view ^{99m}Tc SC liver–spleen scan. Large focal defects due to metastatic colon cancer involving right lobe of liver.

Middle: Comparable three-view initial ^{99m}Tc-MAA perfusion study. There is minimal perfusion of the right lobe of liver. The perfusion defect is considerably larger than tumor involvement seen on similar views on ^{99m}Tc SC. The large central hot spot represents hepatic arterial thrombosis (*white arrow*). Gastric extrahepatic perfusion is noted (*black arrowheads*).

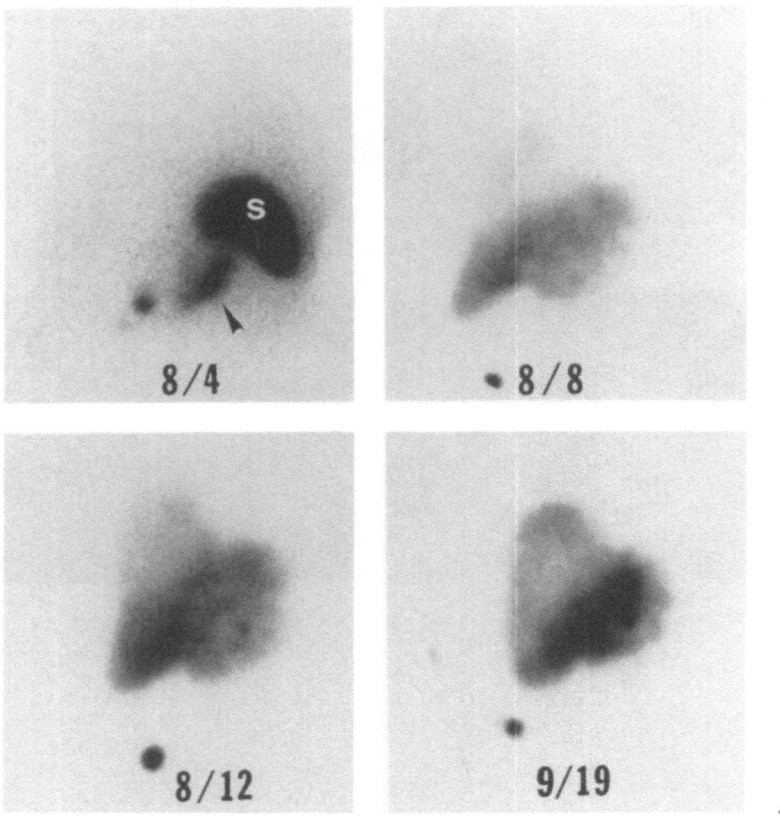

A

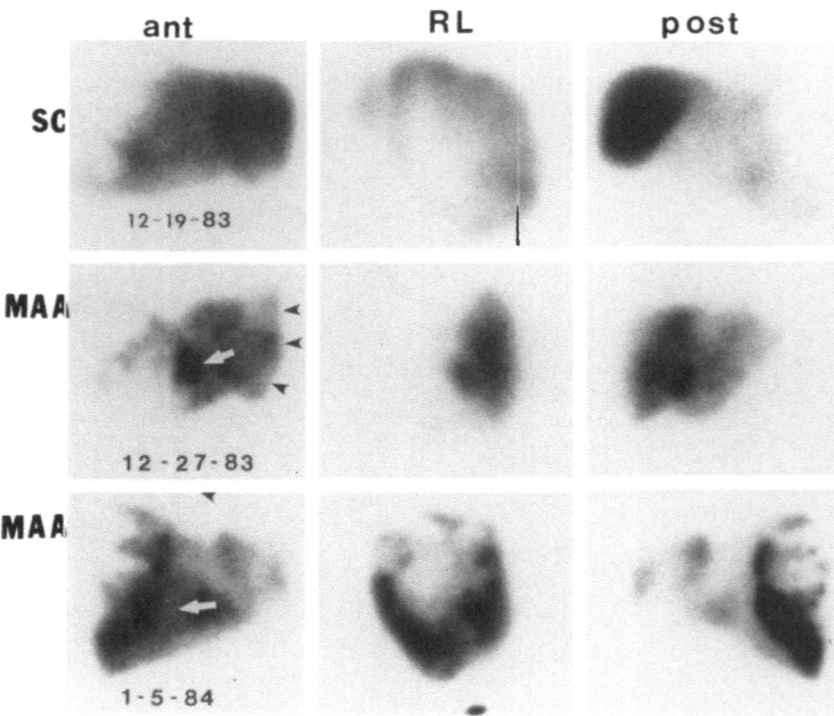

B-1

B-2

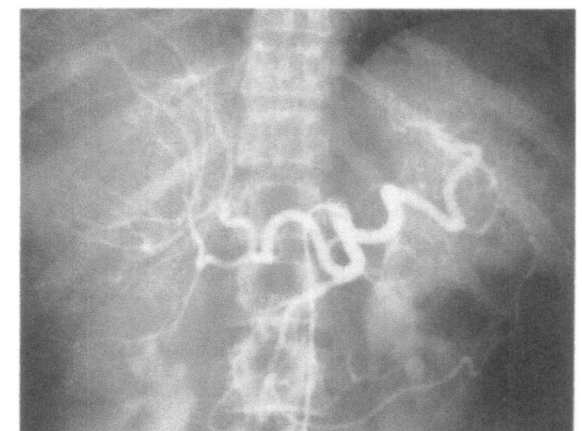

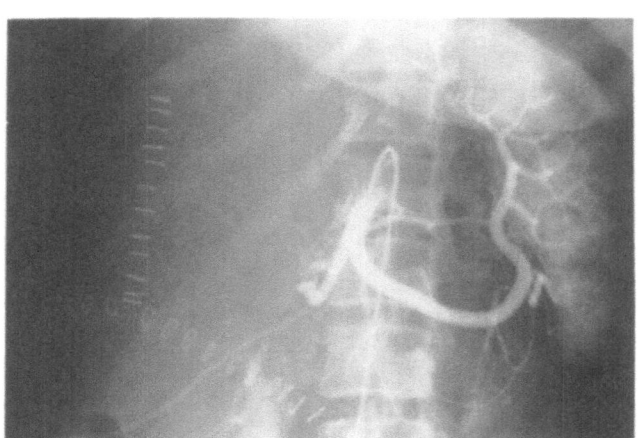

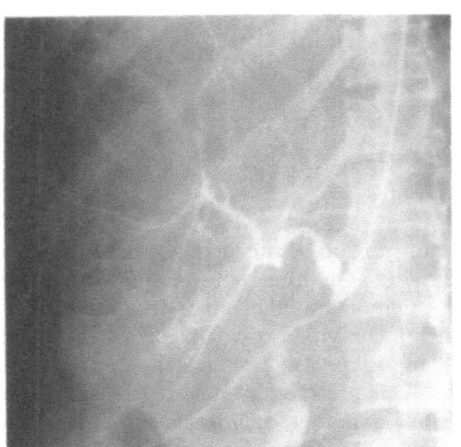

B-1 *Bottom:* Poststreptokinase ^{99m}Tc-MAA study. There is reperfusion of the right lobe best seen on the right lateral and posterior views. Note the hyperperfused peripheral rim and central necrotic core of this large tumor. The hot spot has resolved (*white arrow*).

(Reprinted from Ziessman HA, Juni JE, Gyves JW, Brady TM, Ensminger WD. Thrombosis and streptokinase lysis during hepatic intraaraterial chemotehrapy: the value of perfusion scintigraphy. *AJR.* 1985;144:1067–1068. With permission.)

B-2: *Top:* Selective arteriogram before surgery. Baseline hepatic vascular anatomy, and large mass in the right lobe of liver. *Middle:* Oblique view of selective hepatic arteriorgram 7 days after operation. Occlusion of proper hepatic artery. *Bottom:* Arteriogram 48 hr after streptokinase infusion. Proper hepatic and right hepatic arteries are patent, but the left hepatic artery remains occluded.

(Reprinted from Ziessman HA, Juni JE, Gyves JW, Brady TM, Ensminger WD. Thrombosis and streptokinase lysis during hepatic intraaraterial chemotherapy: the value of perfusion scintigraphy. *AJR.* 1985; 144:1067–1068. With permission.)

140   Harvey A. Ziessman

Figure 4.26. Changes in perfusion pattern induced by therapeutic interventions.

Adjunctive forms of therapy may help improve the delivery of chemotherapy to the liver. Hepatic arterial perfusion studies are useful for evaluating changes in blood flood as a result of these new therapies.

A: Degradable starch microsperes. The initial anterior digital image (*left*) after the infusion of 2 mCi of ^{99m}Tc-MAA demonstrates perfusion predominantly to the right lobe of the liver. Without moving the patient, a second infusion of 2 mCi of ^{99m}Tc-MAA in suspension with 36 million starch particles (3 ml) shows increasing perfusion to the left lobe of the liver (*middle*). Computer subtraction of the first image from the second (*right*) demonstrates that perfusion had been predominantly diverted from the right lobe to the left lobe, stomach (*arrow*) and spleen (S).

Comment: This patient received adjunctive intraarterial chemotherapy with degradable starch microspheres. They were infused in suspension with the chemotherapeutic drug. The starch microspheres temporarily decrease liver blood flow, resulting in a more prolonged exposure of the tumor to the drug and less sytemic exposure and toxicity. To assure that no change in the pattern of perfusion would occur as a result, a ^{99m}Tc-MAA study was performed and acquired on computer.

(Reprinted from Ziessman HA, Gyves JW, Juni JE, et al. Atlas of hepatic arterial perfusion scintigraphy. *Clin Nucl Med*. 1985; 10:675–681. With permission.)

B: Change in tumor/nontumor perfusion ratio with intraarterial vasoconstrictor therapy. These are posterior views of the right lobe of the liver. Note the large hyperperfused rim of tumor activity and hypoperfused central core on the baseline image. After the baseline 1 mCi ^{99m}Tc-MAA study (*upper left*), norepinephrine was infused at three sequentially increasing dose rates, each time followed by another 1 mCi ^{99m}Tc-MAA infusion study. The sequential computer-subtracted images ("net" 1,2,3) demonstrate a changing pattern of perfusion at higher dose rates. In the last image (*bottom right*), improved tumor-to-nontumor perfusion is seen in the inferior aspect of the right lobe. However, there has been undesirable shunting of perfusion away from the superior portion of the large tumor nodule.

Comment: Quantiative SPECT studies have demonstrated tumor-to-nontumor perfusion activity ratios of 2–8 : 1 (mean 3 : 1) in colon cancer metastatic to the liver.[46] This preferential tumor perfusion is a major advantage of intraarterial chemotherapy. Adjunctive intraarterial chemotherapy designed to improve this tumor-to-nontumor ratio could potentially increase drug delivery to the tumor and improve the response rate. Since tumor microvasculature usually lacks a smooth muscle layer in contrast to nontumor tissue, it does not respond normally to vasoactive agents. Therefore, vasoconstrictors have the potential to increase the tumor-to-nontumor blood flow ratio by shunting blood from uninvolved normal liver to tumor, thereby improving chemotherapeutic delivery to the tumor. ^{99m}Tc-MAA perfusion studies can qualitatively and quantitatively evaluate the effectiveness of a particular vasoconstrictor at various dose rates in an individual patient. In this example, norepinephrine is used as a vasoconstrictor. However, other vasoconstrictors are being studied that might work better at the capillary level (e.g., angiotensin II). Norepinephrine seems to result in large vessel vasoconstriction, frequently producing extrahepatic shunting.

(Reprinted from Ziessman HA, Gyves JW, Juni JE, et al. Atlas of hepatic arterial perfusion scintigraphy. *Clin Nucl Med*. 1985; 10:675–681. With permission.)

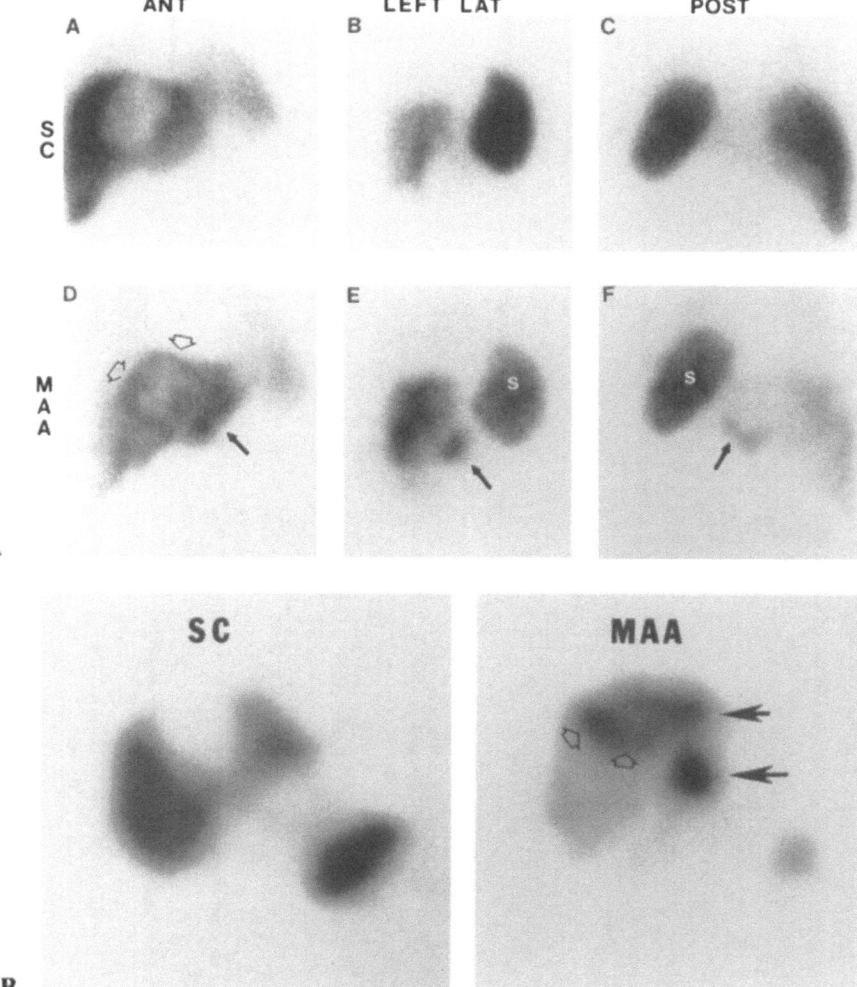

ANT   LEFT LAT   POST

Figure 4.27. Utility of SPECT.

Suspicious gastric perfusion on planar study confirmed with SPECT.

A: Planar perfusion study. The ^{99m}Tc SC study (*above*) shows a large tumor in the midportion of the liver. The ^{99m}Tc-MAA intraarterial pefusion study (*below*) shows a large tumor nodule (*arrowheads*), which corresponds to large focal defect seen on the ^{99m}Tc SC liver–spleen scan. Note the hyperperfused peripheral rim and hypoperfused center of the tumor. Perfusion to spleen (S) is seen on the left lateral view and posterior projections. Whenever the spleen is perfused, gastric perfusion is usually present as well, if carefully looked for. Note increased activity along inferior medial aspect of left lobe of liver (D, *arrow*) that appears to correspond to uninvolved liver on ^{99m}Tc SC scan (A). This can also be seen in posterior view (F, *arrow*) and left lateral view behind left lobe of liver and anterior to the spleen (E, *arrow*). This suggests extrahepatic perfusion to stomach.

(Reprinted from Ziessman HA, Wahl RL, Juni JE, et al. The utility of SPECT for ^{99m}Tc-MAA hepatic arterial perfusion scintigraphy. *AJR*. 1985;145:747–751. With permission.)

B: This SPECT study on the patient in Fig. 27 A. confirms gastric perfusion suspected on the planar study. Paired images are of transaxial tomographic slices, with ^{99m}Tc SC liver–spleen scan on left and comparable cross-sectional ^{99m}Tc-MAA cut on right. The SPECT ^{99m}Tc SC scan shows a large focal tumor defect seen on the planar study. On the corresponding ^{99m}Tc-MAA section, note the hyperperfused rim of tumor (*arrowheads*) and hypoperfused center. Extrahepatic perfusion to the stomach is clearly seen medial and posterior to left lobe (*arrows*).

(Reprinted from Ziessman HA, Wahl RL, Juni JE, et al. The utility of SPECT for ^{99m}Tc-MAA hepatic arterial perfusion scintigraphy. *AJR*. 1985;145:747–751. With permission.)

Comment: SPECT is able to depict the three-dimensional distribution of perfusion, separate out overlying activity, and improve contrast resolution. It has been found useful in evaluating the extent of hepatic perfusion and in determining the presence or absence of extrahepatic perfusion.[45]

Figure 4.28. Utility of SPECT, example 2.

A: Planar perfusion study. *Above:* ^{99m}Tc-MAA scans in anterior, left lateral, and posterior views. *Below:* [99m]Tc SC scans in the same views. There is no definite gastric extrahepatic perfusion. The ^{99m}Tc-MAA study shows complete pefusion to both lobes of liver and slight splenic (*closed arrowheads*) perfusion. Note radioactivity inferior and posterior to the left lobe of liver in left lateral projection (*open arrowhead*). In addition, close inspection of the anterior view sugggests subtle perfusion just medial to left lobe (*open arrowhead*) when compared with the ^{99m}Tc SC scan. This is suggestive of gastric perfusion, but is by no means certain in this planar study.
(Reprinted from Ziessman HA, Wahl RL, Juni JE, et al. The utility of SPECT for ^{99m}Tc-MAA hepatic arterial perfusion scintigraphy. *AJR.* 1985;145:747–751. With permission.)

B: SPECT clearly demonstrates gastric perfusion, which could easily be overlooked on planar study. Transaxial ^{99m}Tc SC slices (*left*) and ^{99m}Tc-MAA cut at comparable level (*right*). ^{99m}Tc-MAA scan shows large metastatic tumor nodule with hyperperfused rim and hypoperfused center, which corresponds to a large focal defect on ^{99m}Tc SC scan. Medial and posterior to tumor nodule is extrahepatic perfusion to stomach (*arrowhead*). Definite splenic perfusion is not seen in this section.
(Reprinted from Ziessman HA, Wahl RL, Juni JE, et al. The utility of SPECT for ^{99m}Tc-MAA hepatic arterial perfusion scintigraphy. *AJR.* 1985;145:747–751. With permission.)

C: CT study with oral contrast in stomach anatomically confirms that the uptake seen on the SPECT study in **B** is gastric perfusion.
(Reprinted from Ziessman HA, Wahl RL, Juni JE, et al. The utility of SPECT for ^{99m}Tc-MAA hepatic arterial perfusion scintigraphy. *AJR.* 1985;145:747–751. With permission.)

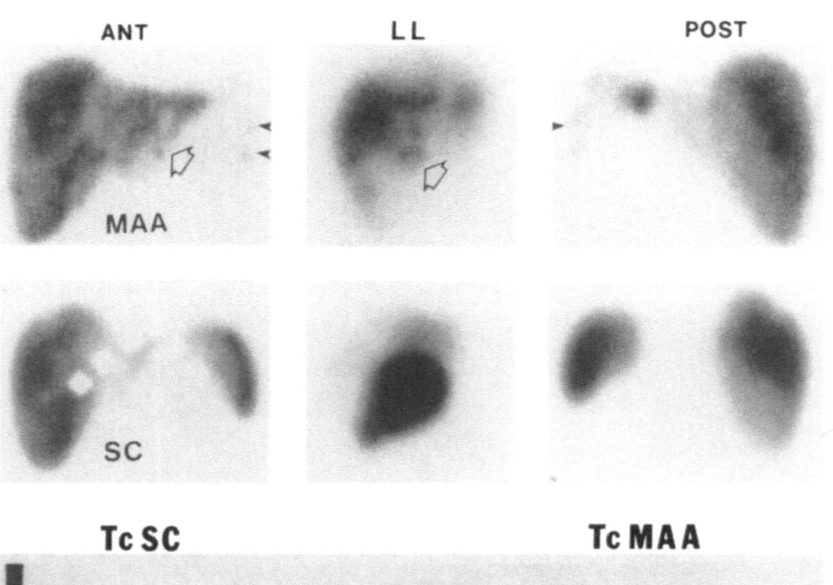

A

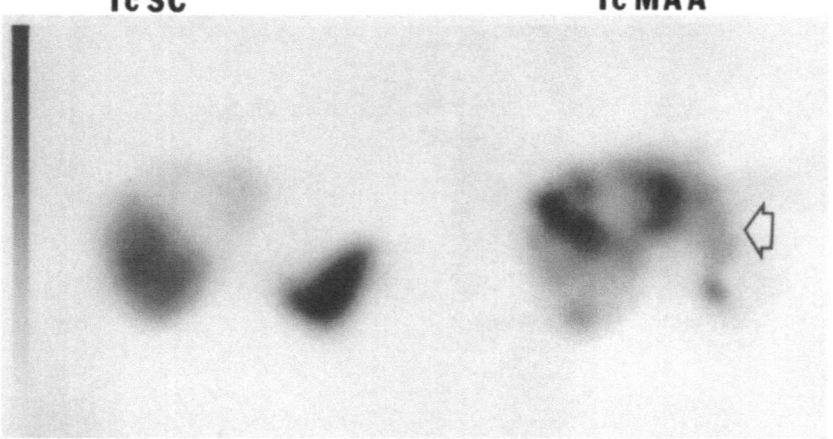

B

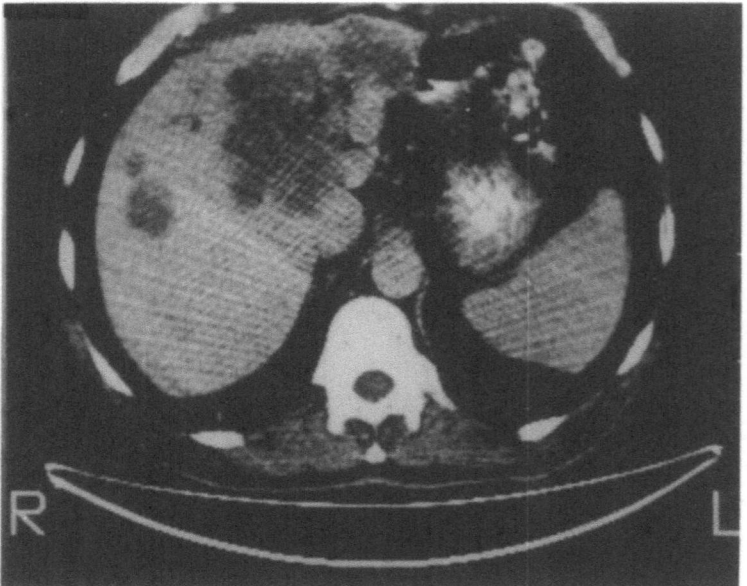

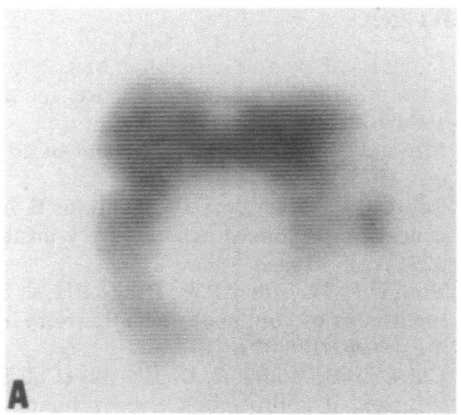

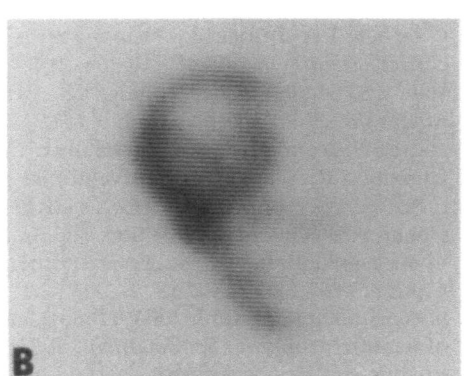

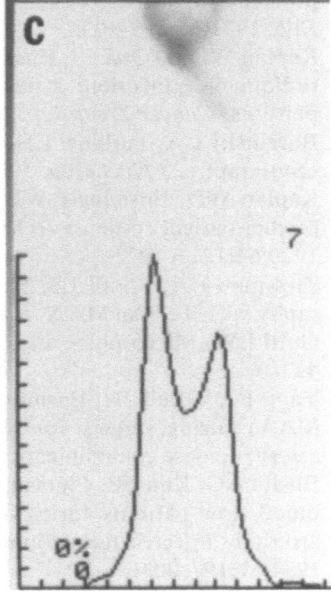

Figure 4.29. Quantitative SPECT.

A: Selected coronal tomographic section on ⁹⁹ᵐTc SC study showing a large focal defect in a patient with proven metastatic colon cancer.

B: Comparable ⁹⁹ᵐTc-MAA coronal slice on ⁹⁹ᵐTc-MAA study showing a rim of increased activity at periphery of metastatic nodule (*arrow*).

C: Profile histogram through the center of the tumor nodule using the coronal section. Counts are displayed on vertical axis and pixels on horizontal axis. The ⁹⁹ᵐTc-MAA tumor-to-liver count ratio was determined by comparing maximal count density of tumor nodule with the count density in adjacent uninvolved liver tissue.

Comment: In a reported series, the tumor-to-nontumor perfusion ratio had a mean of 3.3 (range of 2–8) in colon cancer metastatic to the liver.[46] Obviously, the higher the tumor-to-nontumor ratio, the greater the advantage of intraarterial chemotherapy.

References

1. Stagg RJ, Lewis BJ, Friedman MA, Ignoffo RJ, Hohn DC. Hepatic arterial chemotherapy for colorectal cancer metastatic to the liver. *Ann Intern Med.* 1984;100:736–743.

2. Moertel CG. Clinical management of advanced gastroinestinal cancer. *Cancer.* 1975;36:675–682.

3. Sullivan RD, Norcross JW, Watkins E Jr. Chemotherapy of metastatic liver cancer by prolonged hepatic-artery infusion. *N Engl J Med.* 1964;270:321–327.

4. Brennan MJ, Talley RW, Drake EH, et al. 5-Fluorouracil treatment of liver metastases by continuous hepatic artery infusion via Cournand catheter. *Ann Surg.* 1963;158:405–419.

5. Clarkson B, Young C, Dierick W, et al. Effects of continuous hepatic artery infusion of antimetabolites on primary and metastatic cancer of the liver. *Cancer.* 1962;15:472–488.

6. Oberfield RA. Intraarterial hepatic infusion chemotherapy in metastatic liver cancer. *Semin Oncol.* 1983;10:206–214.

7. Patt YZ, Mavligit GM, Chuang VP. Percutaneous hepatic arterial infusion of mitomycin C and Floxuridine (FUDR): an effective treatment for metastatic colorectal carcinoma in the liver. *Cancer.* 1980;46:261–265.

8. Chuang VP, Wallace S. Interventional approaches to hepatic tumor treatment. *Semin Roentgenol.* 1983;XVIII:127–135.

9. Cohen AM, Greefield A, Wood WC, et al. Treatment of hepatic metastases by transaxillary hepatic artery chemotherapy using an implanted drug pump. *Cancer.* 1983;51:2013–2019.

10. Niederhuber JE, Ensminger WD. Surgical considerations in the management of hepatic neoplasia. *Semin Oncol.* 1983;10:135–147.

11. Ensminger W, Niederhuber J, Dakhil S, Thrall J, Wheeler R. Totally implanted drug delivery system for hepatic arterial chemotherapy. *Cancer Treat Rep.* 1981;65:393–400.

12. Kaplan WD, D'Orsi CJ, Ensminger WD, Smith EH, Levin DC. Intra-arterial radionuclide infusion: a new technique to assess chemotherapy perfusion patterns. *Cancer Treat Rep.* 1978;62:699–703.

13. Borzutsky CA, Turbiner EH. The predictive value of hepatic artery perfusion scintigraphy. *J Nucl Med.* 1985;26:1153–1156.

14. Kaplan WD, Ensminger WD, Come SE, et al. Radionuclide angiograpy to predict patient response to hepatic artery chemotherapy. *Cancer Treat Rep.* 1980;64:1217–1222.

15. Ziessman HA, Thrall JH, Yang PJ, et al. Hepatic arterial perfusion scintigraphy with Tc-99m MAA. *Radiology.* 1984;152:167–172.

16. Gotti EW. Microsphere angiography of the liver. *J Nucl Med.* 1978;19:433–434.

17. Yang PJ, Thrall JH, Ensminger WD, et al. Perfusion scintigraphy (Tc-99m MAA) during surgery for placement of chemotherapy catheter in hepatic artery: concise communication. *J Nucl Med.* 1982;23:1066–1069.

18. Bledin AG, Kim EE, Chuang VP, Wallace S, Haynie TP. Changes of arterial blood flow patterns during infusion chemotherapy, as monitored by intra-arterially injected technetium-99m macroaggregated albumin. *Br J Radiol.* 1984;57:197–203.

19. Clouse ME, Ryan AR, Oberfield RA, McCaffrey JA. Complications of long-term transbrachial hepatic arterial infusion chemotherapy. *AJR.* 1977;129:799–803.

20. Kemeny N, Cohen A, Bertino JR, et al. Continuous intrahepatic infusion of floxuridine and leucovorin through an implantable pump for the treatment of hepatic metastases from colorectal carcinoma. *Cancer.* 1989;65:2446–2450.

21. Kemeny N, Daly J, Reichman B, et al. Intrahepatic or systemic infusion of fluorodeoxyuridine in patients with liver metastases from colorectal carcinoma. *Ann Intern Med.* 1987;107:459–465.
22. Hohn D, Stagg R, Friedman M, et al. The NCOG randomized trial of intravenous vs hepatic arterial FUDR for colorectal cancer metastatic to the liver. *Proc Am Soc Clin Oncol.* 1987;6:85. Abstract.
23. Kemeny N, Goldberg D, Beatty JD, et al. Results of a prospective randomized trial of continuous regional chemotherapy and hepatic resection as treatment of hepatic metastases from colorectal primaries. *Cancer.* 1968;57:492–498.
24. O'Connell M, Mailliard J, Martin J, et al. A controlled trial of regional intra-arterial FUDR versus systemic 5 FU for the treatment of metastatic colorectal cancer confined to the liver. *Proc Am Soc Clin Oncol.* 1989;8:98. Abstract.
25. Safi F, Bittner R, Roscher R, et al. Regional chemotherapy for hepatic metastases of colorectal carcinoma (continuous intra-arterial versus continuous intraarterial/intravenous therapy). *Cancer.* 1989;64:379–387.
26. Chen H-S, Gross JF. Intra-arterial infusion of anticancer drugs: theoretic aspects of drug delivery and review of responses. *Cancer Treat Rep.* 1980; 64:31–40.
27. Ensminger WE, Gyves JW. Clinical pharmacology of hepatic arterial chemotherapy. *Semin Oncol.* 1983;10:176–181.
28. Dakhil S, Ensminger WD, Cho K, et al. Improved regional selectivity of hepatic arterial BCNU with degradable microspheres. *Cancer.* 1982;50:631–635.
29. Ziessman HA, Thrall JH, Gyves JW, et al. Quantitative hepatic arterial perfusion scintigraphy and starch microspheres in cancer chemotherapy. *J Nucl Med.* 1983;24:871–875.
30. Zlotecki R, Gyves J, Ensminger W. Epinephrine induced shunting of blood flow to rabbit hepatic VX-2 implants. *Proc Am Assoc Cancer Res.* 1985;26:1392.
31. Ziessman HA, Forastiere Wheeler RH, et al. The use of a vasoconstrictor to improve the tumor to non-tumor blood flow ratio in intraarterial chemotherapy of head and neck cancer: work in progress. *Nucl Med Commun.* 1985;6:777–786.
32. Ziessman HA, Gyves JW, Juni JE, et al. Atlas of hepatic arterial perfusion scintigraphy. *Clin Nucl Med.* 1985;10:675–681.
33. Noguchi S, Miyauchi K, Nishizawa Y, et al. Augmentation of anticancer effect with angiotensin II in intraarterial infusion chemotherapy for breast carcinoma. *Cancer.* 1988;62:467–473.
34. Wollner IS, Knutsen CA, Ullrich KA, et al. Effects of hepatic arterial yttrium-90 microsphere administration alone and combined with regional bromodeoxyuridine infusion in dogs. *Cancer Res.* 1987;47:3285–3290.
35. Houle S, Yip T-CK, Shepherd FA, et al. Hepatocellular carcinoma: pilot trial of treatment with Y-90 microspheres. *Radiology.* 1989;172: 857–860.
36. Wollner I, Knutson C, Smith P, et al. Effects of hepatic arterial yttrium-90 glass micropsheres in dogs. *Cancer.* 1988;61:1336–1344.
37. Shibata J, Fujiyama S, Sato T, et al. Hepatic arterial injection chemotherapy with cisplatin suspended in an oily lympographic agent for hepatocellular carcinoma. *Cancer.* 1989;64:1586–1594.
38. Sasaki Y, Imaoka S, Kasugai H, et al. A new approach to chemoembolization therapy for hepatoma using ethiodized oil, cisplatin, and gelatin sponge. *Cancer.* 1987;60:1194–1203.
39. Wahl RL, Ziessman HA, Juni J, Lahti D. Gastric air contrast: useful adjunct to hepatic artery scintigraphy. *Am J Roentgenol.* 1984;143:321–325.
40. Croft BY. Radiation dose estimates for the arterial injection of Tc-99m labeled HSA microspheres and macroaggregated particles. In: Schlafke-

Stelson AT, Watson EE, eds. *Proceedings of the Fourth International Radio-pharmaceutical Dosimetry Symposium*. Oak Ridge, Tenn: U.S. Dept. of Energy and Oak Ridge Associated Universities; 1986:260–266.

41. Thrall JH. Hepatic artery perfusion study. In: Carey JE, Kline RC, Keyes JW Jr, eds. *Manual of Nuclear Medicine Procedures*. 4th ed. Boca Raton, Fla: CRC Press; 1983:97–99.

42. Michels NA. Newer antomy of the liver and its variant blood supply and collateral circulation. *Am J Surg*. 1966;112:337–347.

43. Chuang VP. Hepatic tumor angiography: a subject review. *Radiology*. 1983; 148:633–639.

44. Kaplan WD, Come SE, Takuorian RW, et al. Pulmonary uptake of technetium-99m macroaggregated albumin: a predictor of gastrointestinal toxicity during hepatic artery perfusion. *J Clin Oncol*. 1984;2:1266–1269.

45. Ziessman HA, Wahl RL, Juni JE, et al. The utility of SPECT for hepatic arterial perfusion scintigraphy. *Am J Roentgenol*. 1985;145:747–751.

46. Gyves JW, Ziessman HA, Ensminger WD, et al. Definition of hepatic tumor microcirculation by single photon emission computed tomography (SPECT). *J Nucl Med*. 1984;25:972–977.

CHAPTER 5

Atlas of Peritoneoscintigraphy

Douglas Van Nostrand and Jay Anderson

The peritoneal cavity may be studied by both radiographic techniques (peritoneography) and radionuclide scintigraphy (peritoneoscintigraphy). Although the radiographic studies offer higher resolution images, the radionuclide studies offer a simple procedure that, in the appropriate situation, answers the clinical question while using a potentially less toxic and more physiologic intraperitoneal agent.

Although neither peritoneography or peritoneoscintigraphy have found widespread clinical utility, peritoneoscintigraphy has been used for many years (e.g., to document the distribution of intraperitoneal radioactive or nonradioactive therapeutic agents before their injection). It has found new uses (e.g., evaluation of complications of continuous ambulatory peritoneal dialysis), and has potential for more uses (e.g., diagnostic intraperitoneal radionuclide monoclonal antibody imaging). Because of its expanding clinical utility, an understanding of peritoneoscintigraphy is warranted. This chapter describes the technique, discusses the clinical indications, and presents an atlas of peritoneoscintigraphy.

Clinical Utility

The clinical indications for peritoneoscintigraphy are as follows:

1. Documentation of the correct placement of peritoneal catheter before subsequent injection of diagnostic or therapeutic radiopharmaceuticals or drugs.

   1. Pre-P-32 or radiolabeled monoclonal antibodies
   2. Prechemotherapeutic agents for the treatment of ascites
   3. LeVeen shunt evaluation.

P-32 chromic phosphate may be used as adjunctive therapy for ovarian carcinoma or for treatment of malignant ascites. In order to help maximize the therapeutic benefit and to minimize complications, it is important that the P-32 chromic phosphate be truly injected into the peritoneal cavity and that the radiopharmaceutical distributes well throughout all

---

Disclaimer. The opinions or assertions contained herein are the private views of the authors/editors and are not to be construed as official or as reflecting the views of the Uniformed Services University of Health Sciences, United States Army, or the Department of Defense.

or most of the intraperitoneal space. To help assure that the therapeutic agent will be injected into the peritoneal cavity and will have good distribution within this cavity, peritoneoscintigraphy with such agents as ^{99m}Tc SC is of value. Before administering the nonrecoverable therapeutic agent, one can identify inappropriately placed catheters (e.g., abdominal wall, rectus sheath, or bowel lumen) and relative contraindications (e.g., loculations due to adhesions). Peritoneoscintigraphy may also be of value before the administration of chemotherapeutic agents (e.g., intraperitoneal tetracycline for malignant ascites) for similar reasons.

LeVeen shunts have been of value in the treatment of ascites secondary to liver disease. A LeVeen shunt is a tube with a one-way valve placed into the intraperitoneal space with the other end of the tube placed into the venous system. The LeVeen shunt allows flow of the ascitic fluid back into the venous system, thereby controlling abdominal distension and its complications. However, patients may present with increasing abdominal girth, and to evaluate for a nonfunctioning LeVeen shunt, peritoneoscintigraphy may be of value. By injecting a radiopharmaceutical into the intraperitoneal cavity, one then monitors for the appearance of radioactivity within the liver or lung. The latter will depend on the radiopharmaceutical used. An understanding of peritoneoscintigraphy helps assure that the absence of radioactivity in the liver or lung is not due to an inappropriately placed catheter or loculation within the abdomen.

2. Demonstration of the final distribution of intraperitoneal injected therapeutic radiopharmaceuticals to help predict complications.

1. Post-P-32 therapy
2. Postradiolabeled monoclonal antibody therapy.

To minimize complications of intraperitoneal radiotherapy, pretherapy peritoneoscintigraphy is the most important step, as discussed above. However, after the administration of the radiopharmaceutical, it is recommended that one confirm the final distribution. The final distribution can be different from the pretherapy peritoneoscintigraphy, and recognizable patterns may forewarn the physician of potential complications.

3. Evaluation of complications in patients on continuous ambulatory peritoneal dialysis.

Continuous ambulatory peritoneal dialysis (CAPD) is an alternative to hemodialysis, which is becoming more widely used because of advantages over hemodialysis. Patients on CAPD have no risk of blood-borne viral infections (e.g., Hepatitis B), no need for arteriovenous access, no need to go routinely to dialysis centers, and less expense. In brief, the procedure involves the insertion of a Tenckhoff catheter through the abdominal wall, and each day four dialysate exchanges of approximately 1.5 to 2 liters are performed. The patients may perform the exchanges at home by themselves, and the patients may ambulate and perform routine activities between exchanges.

However, CAPD is not without its complications. The increased intraabdominal pressure associated with CAPD may complicate preexisting or cause new structural defects in the peritoneal wall, which in turn may lead to subcutaneous infiltration, leaks, inguinal/abdominal wall hernias with and without incarceration/strangulation, or diaphragmatic leakage with pleural effusions. Patients undergoing CAPD may also experience retained dialysate. Although many of these complications can be identified and diagnosed by physical exam, frequently physical exam is inconclusive. Peritoneoscintigraphy may be of value in the evaluation of these complications, which present with pain, abdominal swelling, labial edema, exit site leakage, or pleural effusion.

4. Evaluation of pleural effusions of undetermined etiology in patients with a predisposition of pleuroperitoneal communication (e.g., ascites secondary to cirrhosis, CAPD).

In those patients with pleural effusions of undetermined etiology and a predisposition of pleuroperitoneal communication, peritoneoscintigraphy offers a simple, noninvasive, and inexpensive procedure for confirming pleuroperitoneal communication. After the injection of the radiopharmaceutical into the intraperitoneal cavity, one would monitor for the appearance of the radiopharmaceutical in the pleural effusion.

5. Interpretation of intraperitoneal monoclonal antibody imaging.

Radiolabeled monoclonal antibodies are being developed and evaluated for diagnostic imaging and therapy. Several radiolabeled monoclonal antibodies are currently under evaluation for the detection of ovarian and colon carcinoma with injections intraperitoneally (peritoneoscintigraphy). These could have potential value in the early detection of microscopic ovarian or colon carcinoma and potential value in subsequent treatment. When these become available, peritoneoscintigraphy and the understanding of the various normal distributions of the radiopharmaceuticals injected intraperitoneally will become commonplace.

Technique

Peritoneoscintigraphy is performed by imaging either the intraperitoneal injection of a diagnostic radiopharmaceutical or a previously injected therapeutic radiopharmaceutical. The technique varies with the clinical indication and/or agent used. These are discussed below. The various clinical indications are discussed in more detail later in the text and atlas.

Clinical Indication: Before Intraperitoneal Therapy (Pretherapy)

Patient preparation: NPO 4 to 6 hr before injection. Empty urinary bladder immediately before injection.

Paracentesis: The patient is typically supine; however, the patient may be sitting. With sterile technique and local anesthesia, a needle or catheter is introduced into the peritoneal cavity in the midline between the umbilicus and pubis. This location is used because the linea alba is relatively avascular and reduces the risk of hemorrhage. The catheter may also be introduced laterally.

Radiopharmaceutical: Technetium-99m sulfur colloid (^{99m}Tc SC) or technetium-99m macroaggregated albumin (^{99m}Tc-MAA) may be used.

Dose: 3 to 5 millicuries.

Dilution: 20 to 100 cc of normal saline is recommended, followed by 100 to 500 additional cc of normal saline.

Imaging: 140-KEV photopeak. All-purpose parallel-hole collimator. 100,000 counts/image. If the initial image does not confirm appropriate placement of the catheter, one should repeat images for up to 1 hr and the patient should move or ambulate. If acceptable placement of the catheter is confirmed with adequate distribution demonstrated, one can proceed with the planned intraperitoneal therapy.

Clinical Indication: Evaluation of LeVeen Shunt

Patient preparation: NPO 4 to 6 hr before injection. Empty urinary bladder immediately before injection.

Paracentesis: The patient is typically supine; however, the patient may be sitting. With sterile technique and local anesthesia, a needle or catheter is introduced into the peritoneal cavity in the midline between the umbilicus and pubis. This location is used because the linea alba is relatively avascular and reduces the risk of hemorrhage. The catheter may also be introduced laterally.

Radiopharmaceutical: ^{99m}Tc SC or ^{99m}Tc-MAA may be used.

Dose: 3 to 5 millicuries.

Dilution: Because of the typically large amount of ascitic fluid present, no additional fluid is necessary. However, one may use 5 to 10 cc of normal saline as a flush.

Imaging: 140-KEV photopeak. All-purpose collimator. 100,000 counts/image. When ^{99m}Tc SC is used, obtain initial image of the abdomen to assess the adequacy of the injection. After confirmation of an adequate injection, obtain images of the liver to assess for ^{99m}Tc SC, which has been drained from the abdomen via the LeVeen shunt and has been taken up by the Kupffer's cells of the liver. If radioactivity is noted within the liver, then the LeVeen Shunt is patent. If no radioactivity is noted within the liver, delayed images should be obtained at 1, 2, and 3 hr after injection. When using ^{99m}Tc SC, one should be cautious in interpreting radioactivity within the peritoneal cavity distributing normally *around* the liver from radioactivity *within* the liver. To eliminate this problem, ^{99m}Tc-MAA may be used. If ^{99m}Tc-MAA is used, obtain initial image of the abdomen to assess the adequacy injection. After confirmation of an adequate injection, obtain images of the lung to assess for ^{99m}Tc-MAA, which has been drained from the abdomen via the LeVeen shunt and has been lodged in the pulmonary capillaries. If radioactivity is noted within the lungs, then the LeVeen shunt is patent. If no radioactivity is noted within the lung, delayed images should be obtained at 1, 2, and 3 hr after injection.

Clinical Indication: Demonstration of Final Distribution of Therapeutic Radiopharmaceutical P-32 Chromic Phosphate (Bremsstrahlung Imaging)

Patient preparation: none.

Radiopharmaceutical: Typically performed 1 to 4 days after the administration of the therapeutic P-32 chromic phosphate (P-32 CP). Because fixation of the radiopharmaceutical may not completely occur until after 24 hr, earlier imaging, such as on the day of administration, may not reflect the final distribution.

Camera set-up:

Technique 1[a]: medium energy collimator; peak = 81 KEV; window 90%; 100,000 counts over approximately 30 minutes.

Technique 2[2]: high energy collimator; peak = none; window = 75–300 KEV.

Technique 3[3]: high-energy collimator; peak = none; window = 150–450 KEV; approximately 200,000 counts.

Clinical Indication: Evaluation of Complications Secondary to CAPD

Patient preparation: None.

Paracentesis: None.

Radiopharmaceutical: ^{99m}Tc SC. ^{99m}Tc-MAA may also be used, however, Kopecky[4] has reported that ^{99m}Tc SC appears to have less retention in the abdomen. This can maximize the use of post dialysis images for the detection of leakage. ^{99m}Tc albumin colloid has also been used.

Dose: 3 to 5 millicuries.

Dilution: The radiopharmaceutical is injected into a standard 2-liter dialysate bag, thoroughly mixed, and instilled into the peritoneum with the standard exchange volume through a Tenckhoff catheter using sterile technique.[5]

Imaging: 140-KEV peak; All-purpose collimator; 100,000 counts/image; routine images: supine and upright anterior, lateral, and right and left anterior oblique. Include inguinal canals, external genitalia, and lowermost pleural areas. Place marker on catheter exit site and any other appropriate landmarks desired; obtain initial set of images approximately 10 min after instillation of dialysate. The patient should ambulate between initial and delayed images; obtain delayed images approximately 2 to 5 hr after instillation of dialysate; obtain postdrainage images after drainage of all dialysate.

Clinical Indication: Evaluation of Pleuroperitoneal Communication

For evaluation of patients with pleural effusion who are on CAPD, follow the protocol for the clinical indication "CAPD" above. For evaluation of patients with pleural effusion and cirrhosis/ascites, follow the protocol for the clinical indication of "Pretherapy" and perform additional images of the thorax for several hours after injection.

---

[a]Kaplan[1] has demonstrated the energy spectrum from a P-32–filled capillary pipette when obtained through 2.3 cm of scattering medium. The energy spectrum was from 40 KEV to 1.5 MeV with a peak at 40 to 81 KEV (Fig. 5.7). He reported good results using technique 1 above, but image degradation occurred at higher energies, most likely from Compton scattering from surrounding patient tissue. The lead x ray (78 KEV) emanating from the collimator did not appear to degrade the image significantly.

Clinical Indication: Monoclonal Antibody

The imaging protocol will depend not only on the isotope used (e.g., ^{99m}Tc, ^{111}In, or ^{131}I, but also on the kinetics of monoclonal antibody used.

Radiation Absorbed Dose

Although accurate estimates of radiation absorbed doses for peritoneo-scintigraphy are limited because of "a lack of suitable source region in the phantom calculations," the amount of radiation absorbed dose is not expected to be unacceptable for the potential information gained. Estimates of radiation exposure to the peritoneal lining and other organs should be forth coming using a new geometric model by Watson et al.[6]

Observations and Interpretations

As more and more studies are observed, a typical pattern of peritoneal distribution of radioactivity in normal patients emerges. An alteration of this pattern can offer a significant amount of recognizable and clinically useful information.

A diagrammatic representation of the normal intraperitoneal spaces and reflections are shown in Figure 5.1. The spectrum of normal distribution is demonstrated in Figures 5.2 to 5.10. [99m]Tc SC is the radiopharmaceutical used in Figures 5.2 through 5.6, and bremsstrahlung imaging of P-32 is used in Figures 5.8, 5.9, and 5.10. Examples of abnormal distribution of radioactivity in the intraperitoneal cavity are shown in Figures 5.11 through 5.21. [99m]Tc SC is the radiopharmaceutical used in Figures 5.11 through 5.15, and brehmsstrahlung imaging of P-32 is demonstrated in Figures 5.16 and 5.17. The scintigraphic patterns of complications of CAPD are shown in Figures 5.18 through 5.21. The atlas concludes with an example of abnormal intraperitoneal monoclonal antibody imaging.

Acknowledgment. The authors would like to thank Lisa Anderson for her administrative and secretarial support.

Atlas Section

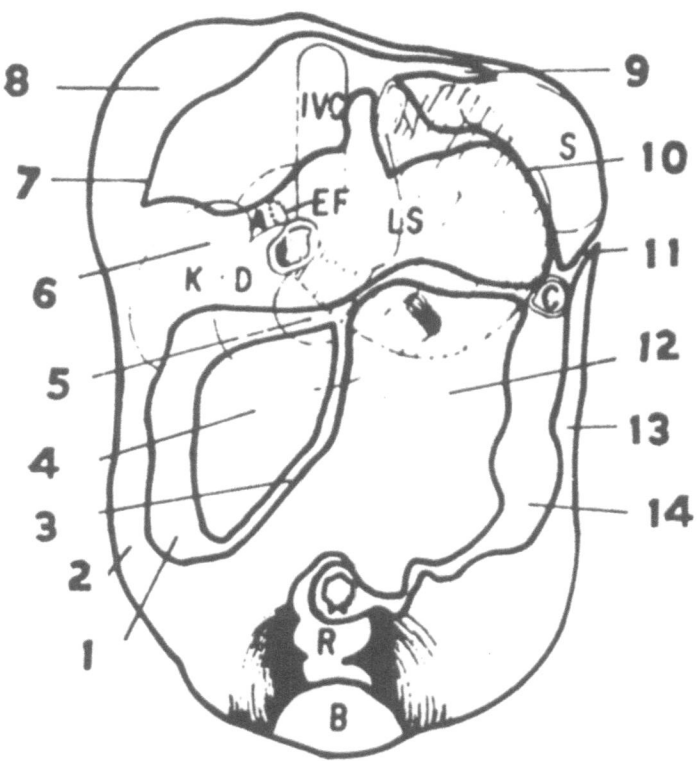

Figure 5.1. Diagram of the posterior peritoneal reflections and recesses.

S, spleen; LS, lesser sac; IVC, inferior vena cava; EF, epiploic foramen; K, right kidney; D, duodenum; A, adrenal gland; C, splenic flexure of colon; R, rectum; B, urinary bladder; 1, attachment of peritoneal reflections of ascending colon; 2, right paracolic gutter; 3, root of mesentery; 4, right infracolic space; 5, root of transverse mesocolon; 6, area of Morrison's pouch; 7, right triangular ligament; 8, right subphrenic space; 9, left triangular ligament; 10, gastrolienal ligament; 11, phrenicocolic ligament; 12, left infracolic space; 13, left paracolic gutter; 14, attachment of peritoneal reflections of descending colon.
(Reproduced with permission from Meyers MA. The spread and localizations of acute intraperitoneal effusions. *Radiology.* 1970; 95:547–554.)[7]

Figure 5.2. Normal peritoneoscintigraphy (^{99m}Tc SC).

This is an anterior view of the abdomen performed shortly after the intraperitoneal administration of ^{99m}Tc SC. This scan demonstrates a typical distribution of radioactivity with the majority of radioactivity in the midabdomen. Radioactivity is also moving cephalad over the liver surface in the patient's right upper quadrant and over the spleen and/or bowel in the patient's left upper quadrant.

(Reproduced with permission from Van Nostrand D, Silberstein EB. Therapeutic use of P-32. In: Freeman LM, Weissman HS, eds. *Nuclear Medicine Annual 1985.* New York: Raven Press; 1985.)[8]

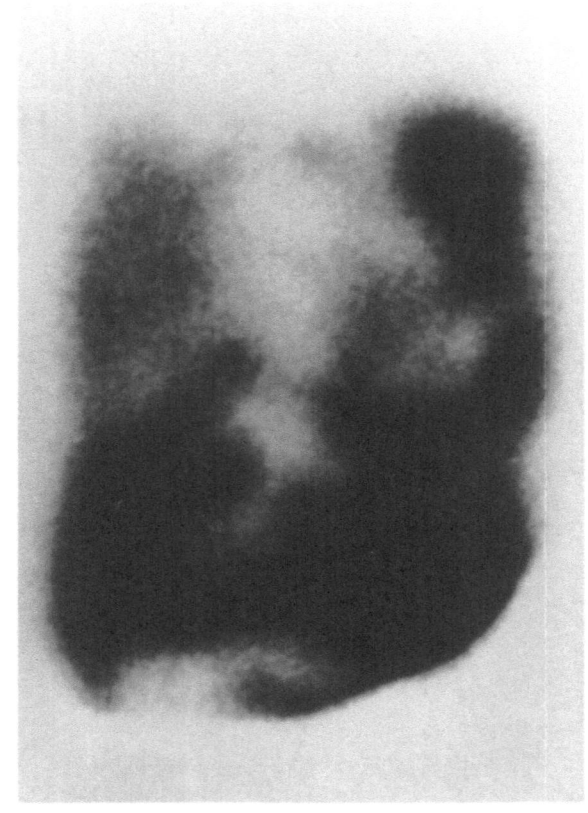

Figure 5.3. Normal peritoneoscintigraphy (^{99m}Tc SC).

This is an anterior view of the abdomen performed approximately 1 hr after the intraperitoneal administration of ^{99m}Tc SC. This scan demonstrates a typical distribution of serpiginous inhomogeneous radioactivity in the abdomen. The circular activity is the peritoneal catheter. Relative to Figure 5.1, more radioactivity is noted extending over the liver and spleen (or bowel) surface in the upper abdomen, and pooling in the cul-de-sac in the lower abdomen is noted.

(Reproduced with permission from Van Nostrand D, Silberstein EB. Therapeutic use of P-32. In: Freeman LM, Weissman HS, eds. *Nuclear Medicine Annual 1985.* New York: Raven Press; 1985.)[8]

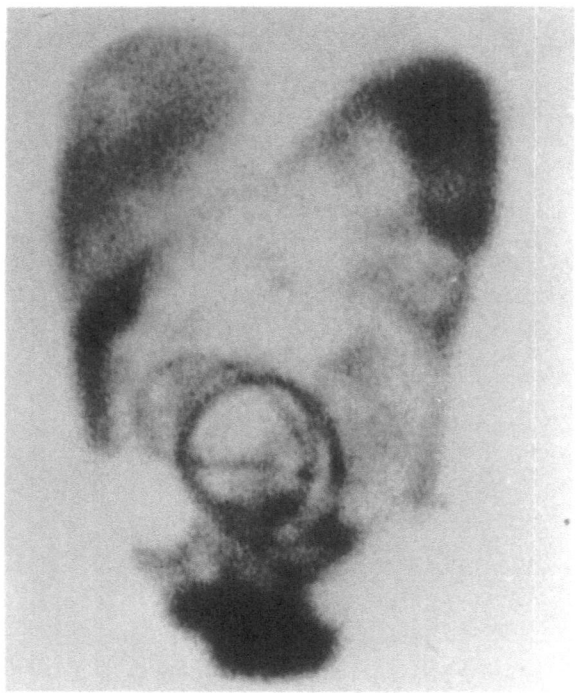

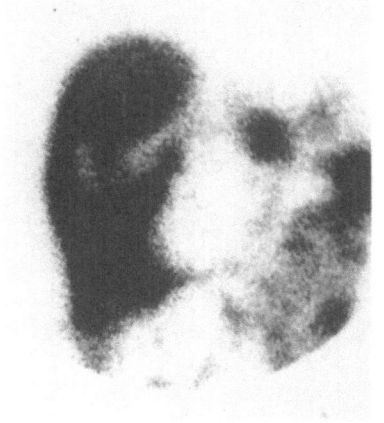
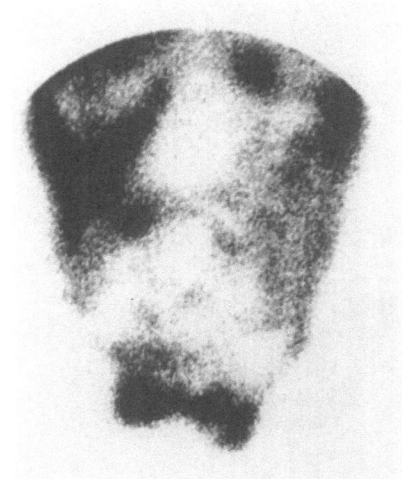

A

B

Figure 5.4. Normal peritoneoscintigraphy (⁹⁹ᵐTc SC).

This scan was performed in the anterior view of the abdomen (**A,** upper abdomen, **B,** lower abdomen and pelvis) within 1 hr after the intraperitoneal administration of ⁹⁹ᵐTc SC. This scan further demonstrates the normal distribution of intraperitoneal radioactivity.

(Reproduced with permission from Neutze J, Van Nostrand D, Major W. Peritoneoscintigraphy in detection of improper placement of peritoneal catheter into bowel lumen prior to chromic phosphate P-32 therapy: a case report. *Clin Nucl Med.* 1985;10:777–779.)[9]

Figure 5.5. Normal peritoneoscintigraphy (⁹⁹ᵐTc SC).

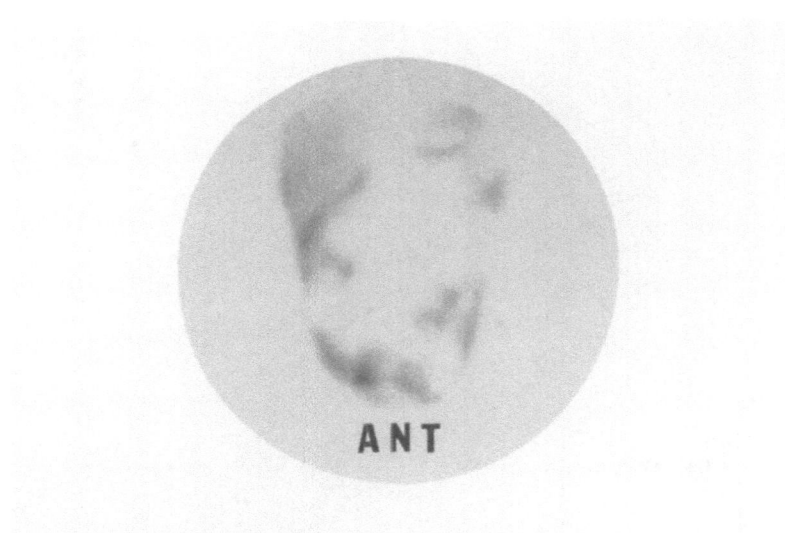

Figure 5.6. Normal peritoneoscintigraphy (^{99m}Tc SC): effect of positioning.

This anterior view, which was performed in the upright (*left image*) and supine (*right image*) position, demonstrates the effect that positioning and gravity may have on distribution. Greater radioactivity is noted in the cul-de-sac in the upright position.

Comment: If good dissemination of the ^{99m}Tc SC is not observed after injection, one should have the patient move into various positions before concluding that distribution is unacceptable for the planned radiotherapy injection. These figures also emphasize that after injection of the radiopharmaceutical or chemotherapy agent, the patient should frequently change positions to help assure good dissemination of the agent in order to maximize the therapeutic effect and to minimize the complications. Recommendations have included a change of position (e.g., supine, prone, reverse trendelenburg, trendelenburg, sitting) every 15 to 20 min; this should continue for 12 to 16 hr after administration of the therapeutic agent.

(Reproduced with permission from Van Nostrand D, Silberstein EB. Therapeutic use of P-32. In: Freeman LM, Weissman HS, eds. *Nuclear Medicine Annual 1985.* New York: Raven Press; 1985.)[8]

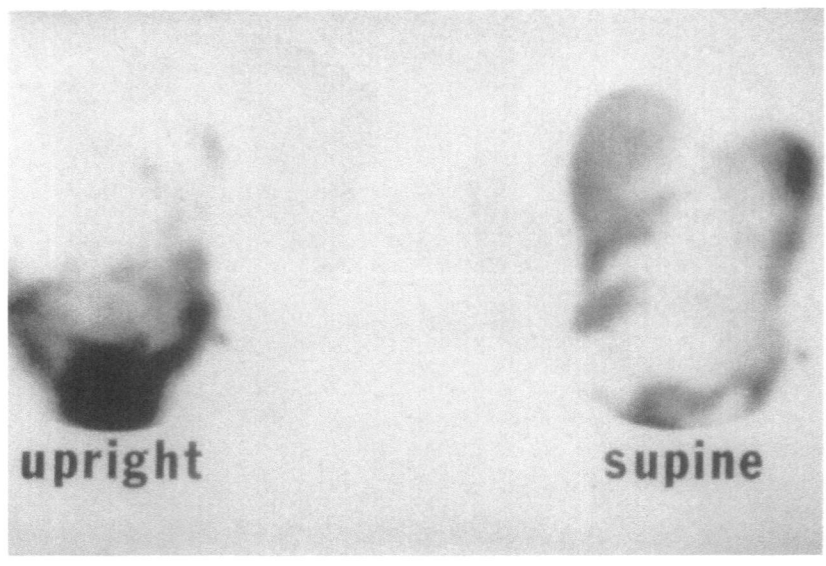
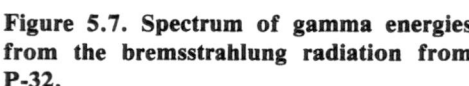

Figure 5.7. Spectrum of gamma energies from the bremsstrahlung radiation from P-32.

P-32 is a pure beta emitter, and during deceleration of the beta particle in a scattering medium such as tissue, bremsstrahlung radiation ("breaking" radiation) is emitted. By imaging this energy, one can obtain images of the final distribution of P-32 CP. This figure demonstrates the energy spectrum of the bremsstrahlung radiation from a P-32 filled capillary pipette when obtained through 2.3 cm of scattering medium.

(Reproduced with permission from Kaplan WD, Zimmerman, RE, Bloomer WD, et al. Therapeutic intraperitoneal P-32: a clinical assessment of the dynamics of distribution. *Radiology.* 1981;138:683.)[1]

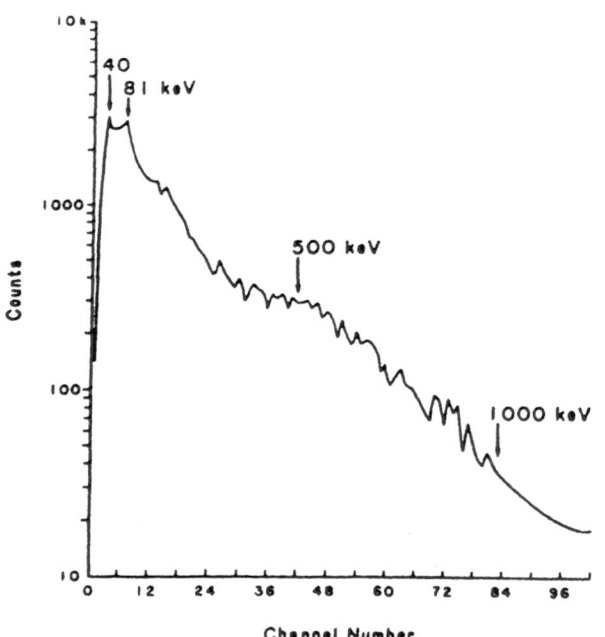

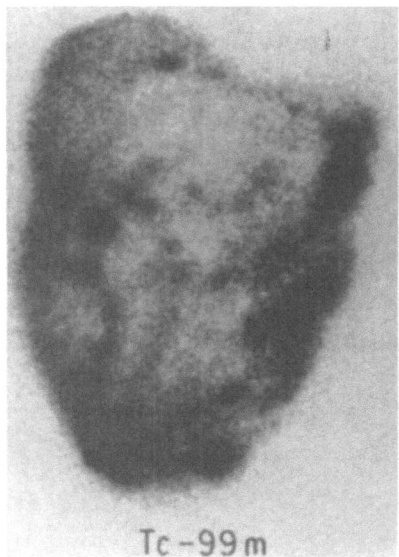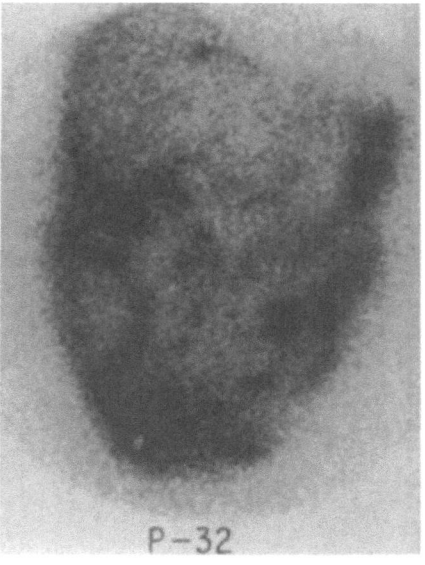

Figure 5.8. Pretherapy: concordance of distribution of ^{99m}Tc SC relative to P-32 CP.

These anterior images of the abdomen are from the same patient and demonstrate good correlation of the intraperitoneal distribution of ^{99m}Tc SC (*left*) with P-32 CP (*right*). The P-32 image was obtained by imaging the bremsstrahlung radiation.
(Reproduced with permission from Sullivan DC, Harris CC, Currie JL, et al. Observations on the intraperitoneal distribution of chromic phosphate (32P) suspension for intraperitoneal therapy. *Radiology*. 1983;146:539.)[3]

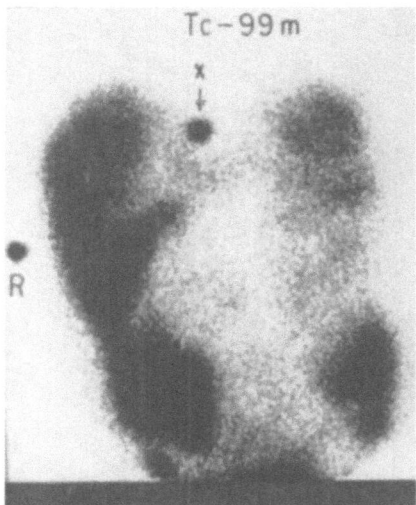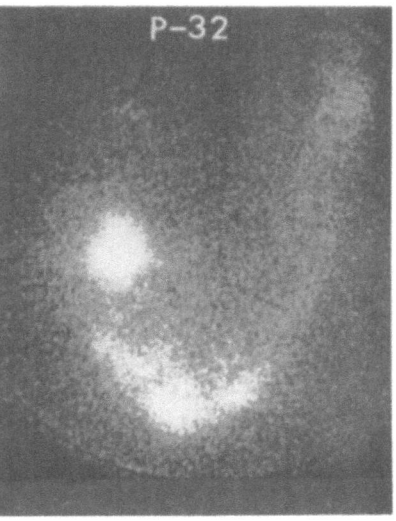

Figure 5.9. Pretherapy: nonconcordance of distribution of ^{99m}Tc SC relative to P-32 CP.

These anterior images of the abdomen are from the same patient and demonstrate a discrepancy between intraperitoneal distribution of ^{99m}Tc SC (*left*) with P-32 CP (*right*). A significant amount of the P-32 is localized in the right lower quadrant at the site of administration. The P-32 image was obtained by imaging the bremsstrahlung radiation.
Comment: The etiology of the focal radioactivity, which was present on the P-32 CP image and not on the ^{99m}Tc SC image, is undetermined, but it is believed to represent slight leakage or adsorption at the injection site.
(Reproduced with permission from Sullivan DC, Harris CC, Currie JL, et al. Observations on the intraperitoneal distribution of chromic phosphate (P-32) suspension for intraperitoneal therapy. *Radiology*. 1983;146:539.)[3]

Figure 5.10. Normal bremsstrahlung imaging: lymph node uptake in the chest.

This is an anterior image of the thorax obtained by imaging the bremsstrahlung radiation for 100,000 counts. Lymph node uptake of the P-32 CP, which was originally injected into the intraperitoneal cavity, is noted.

Comment: When Kaplan evaluated 24 patients after the injection of intraperitoneal P-32, 18 patients demonstrated intrathoracic nodal uptake. The earliest uptake was noted within 24 hr after injection, and the number and uptake intensity of the radioactivity in the lymph nodes remained constant over time. No clinically useful relationship was demonstrated in regard to the presence of intrathoracic lymph node uptake and diaphragmatic tumor.

(Reproduced with permission from Kaplan WD, Zimmerman RE, Bloomer WD, et al. Therapeutic intraperitoneal P-32: a clinical assessment of the dynamics of distribution. *Radiology*. 1981;138:683.)[1]

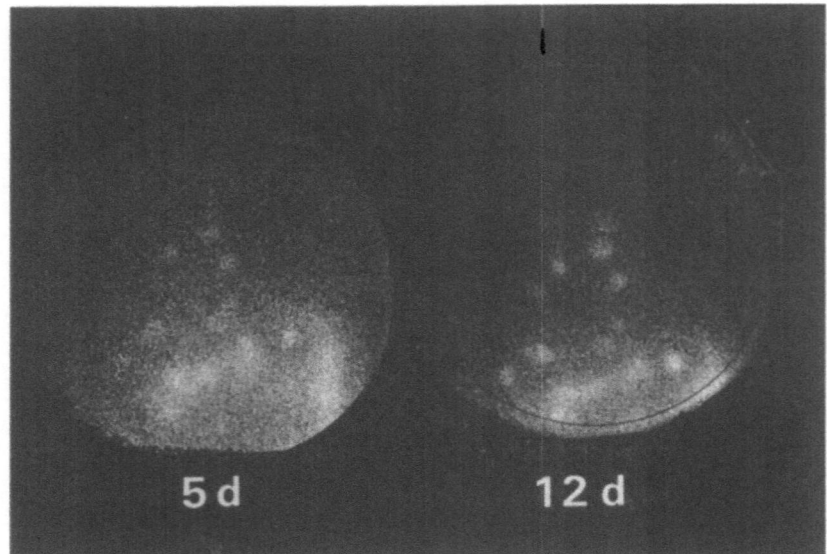

Figure 5.11. Abnormal distribution of 99mTc SC: loculation.

These images, obtained immediately after the injection of 99mTc SC intraperitoneally, demonstrate loculation of the radioactivity.

Comment: Loculations may occur from previous surgery, peritonitis, or neoplasms. The above pattern is a contraindication for using the catheter for the P-32 therapy in the dosage planned. The identification of the loculation will prevent preceding with an essentially useless and potentially harmful therapy. A new or second catheter should be placed and reimaged.

(Reproduced with permission from Van Nostrand D, Silberstein EB. Therapeutic use of P-32. In: Freeman LM, Weissman HS, eds. *Nuclear Medicine Annual 1985.* New York: Raven Press; 1985.)[8]

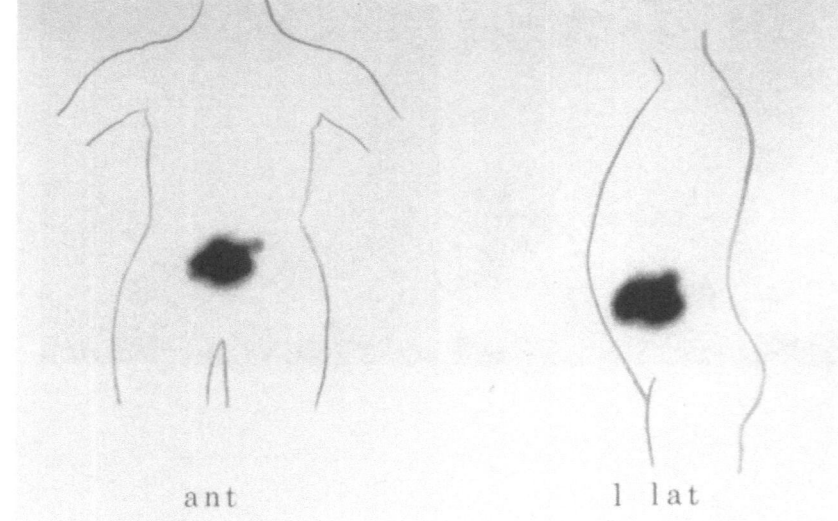

Figure 5.12. Abnormal distribution of ^{99m}**Tc SC: hemiabdomen loculation.**

Both images (*left*) are anterior views performed with ^{99m}Tc SC. **A:** Loculation of the radioactivity in the left hemiabdomen. The line of radioactivity extending laterally (*arrow*) represents the catheter. A second catheter was placed into the right hemiabdomen with additional injection of ^{99m}Tc SC and imaging (**B**). Relatively good but not complete distribution of radioactivity is noted in the right hemiabdomen. The linear radioactivity again represents catheters.

Comment: The above pattern of radioactive distribution is not an absolute contraindication to proceeding with therapy. However, the administered dose injected in each catheter to each area of the abdomen should be reduced. When loculations occur, the decision to proceed with therapy must be made on an individual basis based on such factors as the patient, the pattern of radioactive distribution, and volume of distribution.

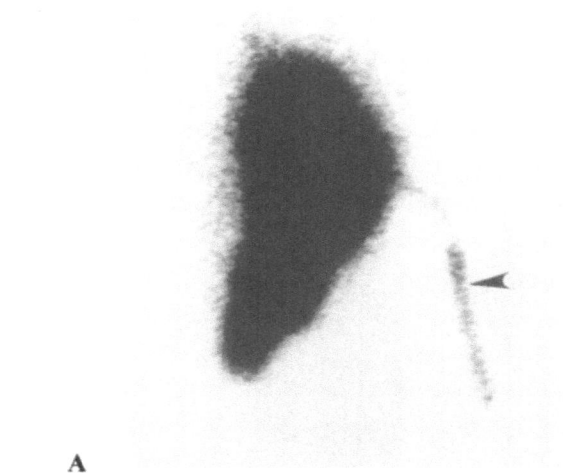

A

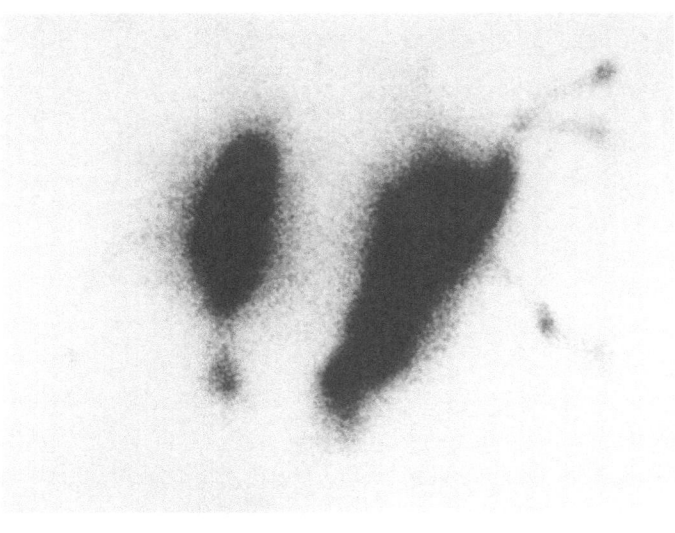

B

Figure 5.13. Abnormal distribution of ^{99m}Tc SC: bowel injection.

Anterior view of the abdomen performed with ^{99m}Tc SC in preparation for P-32 CP therapy. Radioactivity within the injection catheter (*small arrowhead*) is seen, and radioactivity is clearly identified in the small bowel (*curved arrow*), ascending colon (*large arrowhead*), and transverse colon (*small arrows*). Haustra are seen (*arrows*).
Comment: This pattern of radioactivity should be compared with and distinguished from the distribution of a good intraperitoneal injection. In addition to the distinct pattern above, no radioactivity was noted in the cul-de-sac or seen progressing over the liver or spleen. Although this case would appear to be easily identifiable, in the clinical setting and when evaluating distribution only on the persistence scope of the gamma camera, this pattern can be difficult to distinguish from a good injection. Accordingly, evaluation of images on film are recommended.
(Reproduced with permission from Neutze J, Van Nostrand D, Major W. Peritoneoscintigraphy in detection of improper placement of peritoneal catheter into bowel lumen prior to chromic phosphate P-32 therapy: a case report. *Clin Nucl Med.* 1985;10:777–779.)[9]

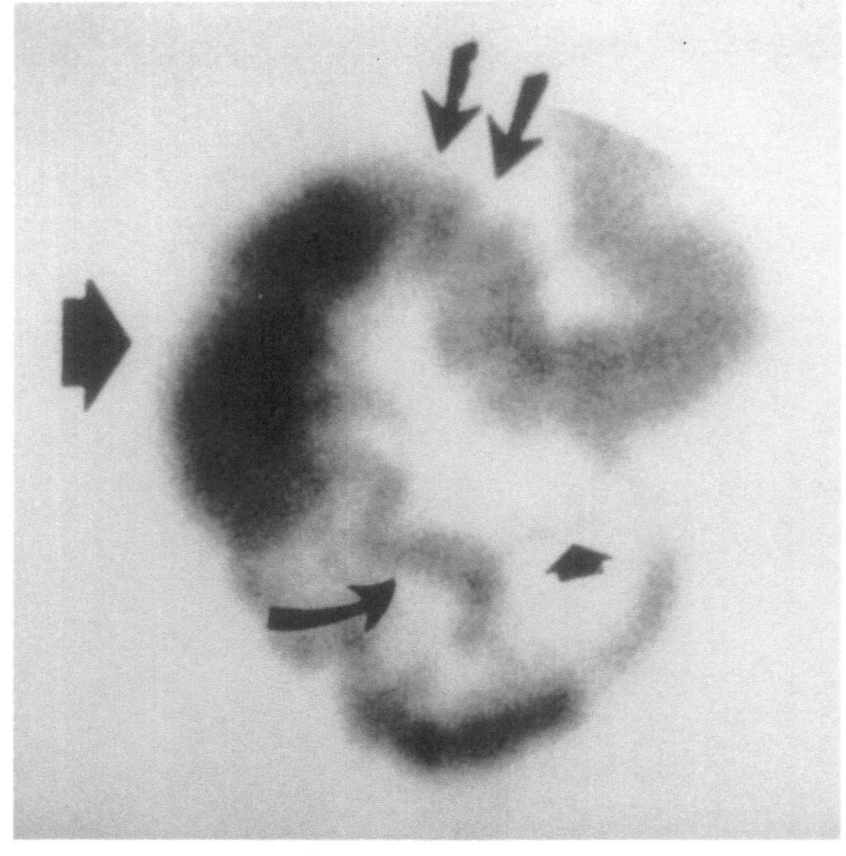

Figure 5.14. Abnormal distribution of ^{99m}Tc SC: pelvic mass.

Parts a, b, and c represent anterior images performed with ^{99m}Tc SC over a 16-month period. They demonstrate diminishing radioactivity in the right lower quadrant, which was secondary to an enlarging tumor.
Comment: Although peritoneoscintigraphy would not be indicated for the detection of abdominal or pelvic masses, the study may reflect the presence of a mass, and one should be aware that masses may affect the distribution of radioactivity or possibly the success of intraperitoneal radiotherapy.
(Reproduced with permission from Arnstein NB, Wahl RL, Cochran M, et al., Adenocarcinoma of the alimentary tract: peritoneal distribution scintigraphy. *Radiology.* 1987;162:439–441.)[10]

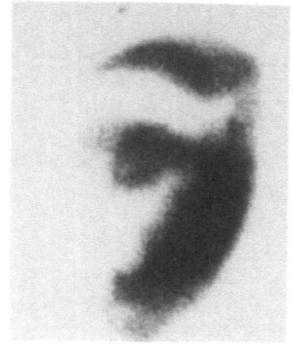
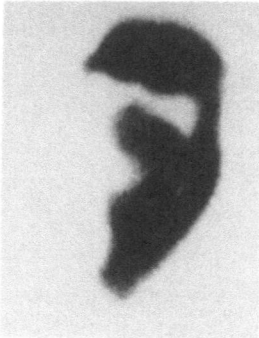

a.   b.   c.

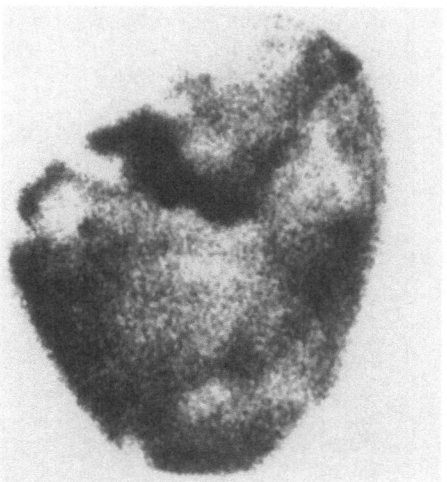

Figure 5.15. Abnormal distribution of ^{99m}Tc SC: liver metastasis.

Anterior view of the abdomen performed with ^{99m}Tc SC demonstrates no activity over the liver, which was despite ambulation of the patient.

Comment: Although not diagnostic, Arnstein has reported that this pattern of absence of activity over the dome of the liver was frequently seen in patients with liver metastasis confirmed on computed tomography. When this pattern is seen, one should confirm that the patient has changed positions according to the established protocol. If this pattern persists and if there is no knowledge of liver metastasis, one should consider further evaluation for such.

(Reproduced with permission from Arnstein NB, Wahl RL, Cochran M, et al. Adenocarcinoma of the alimentary tract: peritoneal distribution scintigraphy. *Radiology.* 1987;162:439–441.)[10]

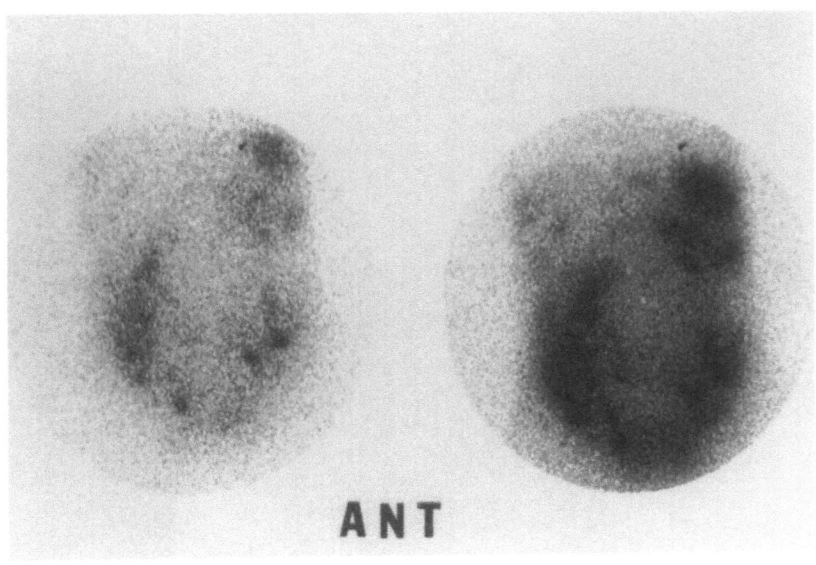

ANT

Figure 5.16. Posttherapy: P-32 CP aggregation.

These two images represent the same anterior view of the abdomen obtained by imaging the bremsstrahlung radiation of P-32 CP with slightly different intensities. In addition to a generalized pattern of P-32 CP distributed throughout the abdomen, focal aggregates of radioactivity are easily identified.

Comment: Kaplan has suggested that these focal aggregates of radioactivity may have short-term and long-term implications. During short-term follow-up and when imaged, all patients with abdomen discomfort invariably had abdominal aggregations of CP corresponding to the site of pain. Kaplan has estimated that the radiation dose may be as much as a fivefold difference between areas of focal aggregates and more uniform distribution, which could result in more long-term morbidity or less therapeutic effect.

Figure 5.17. Posttherapy: P-32 CP aggregation.

Anterior abdominal images obtained by imaging the bremsstrahlung radiation of P-32 CP. The images are of the same patient performed 5 and 21 days after injection of the P-32 CP. The aggregation of radioactivity is focal and fixed.

Comment: Speculatively, the appearance of aggregated radioactivity most likely is a spectrum with the potential of complications related to size and intensity of the aggregated radioactivity. Since the radioactivity may be fixed by 24 hr and since little change appears to occur after 5 days, routine images within this time period may be of prognostic value.

(Reproduced with permission from Kaplan WD, Zimmerman RE, Bloomer WD, et al. Therapeutic intraperitoneal P-32: a clinical assessment of the dynamics of distribution. *Radiology*. 1981;138:683.)[1]

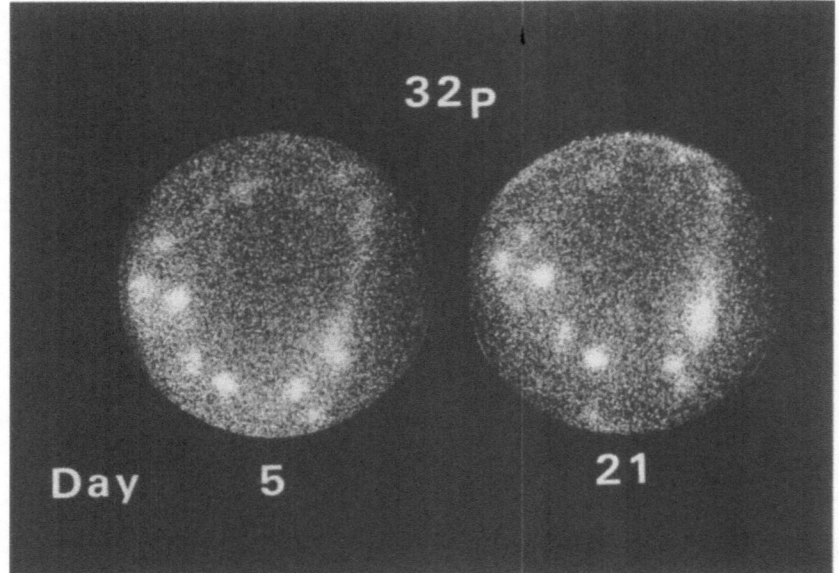

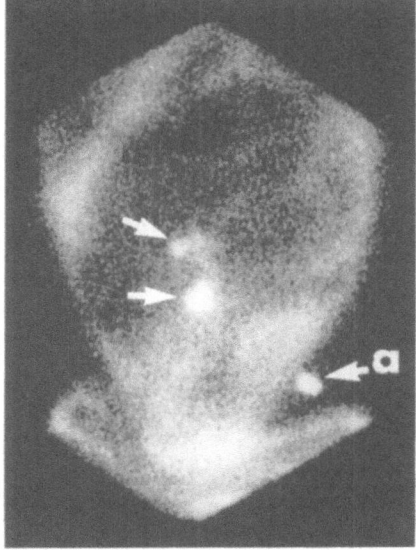

A

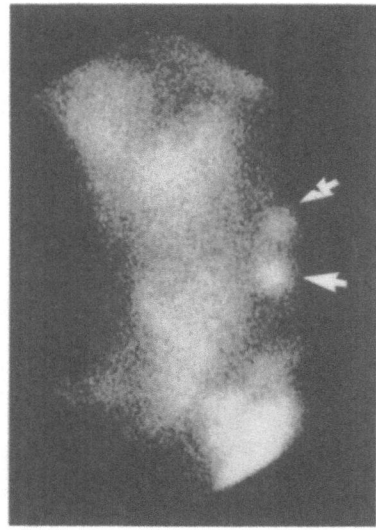

B

Figure 5.18. CAPD: leakage into abdominal wall.

A patient on CAPD presented with diffuse abdominal wall edema. No peritoneal defect was noted on physical exam. Peritoneoscintigraphy was performed 3 hr after the administration of ^{99m}Tc SC in the dialysate. In the anterior (**A**) and right lateral image (**B**), two abnormal areas of radioactive accumulation (*arrows*) are present that represented leakage into the anterior abdominal wall near the umbilicus. A radioactive marker is present at the exit site of the dialysis catheter (*arrow labeled with "a"*).

Comment: CAPD has been used in the management of end-stage renal disease since 1976. However, it is associated with complications such as dialysate leakage, peritonitis, and inadequate ultrafiltration. Peritoneoscintigraphy has been used in the assessment of all of these complications. This case demonstrates the value of the technique not only in identifying the site of leakage of dialysate into the abdominal wall but also the presence of two sites of leakage, which potentially can be repaired surgically.

In addition, the case emphasizes the need to have a good understanding of the normal distribution of intraperitoneal radiopharmaceuticals, to use markers for identifying landmarks (focal areas did not represent catheter site), and to use various modifications of the protocol to maximize the diagnostic yield (e.g., laterals were essential in this case).

(Reproduced with permission from Kopecky RT, Frymoyer PA, Witanowski LS, et. al. Complications of continuous ambulatory peritoneal dialysis: diagnostic value of peritoneal scintigraphy. *Am J Kidney Dis*. 1987;10:123.)[4]

Figure 5.19. CAPD: leakage into patent process vaginalis.

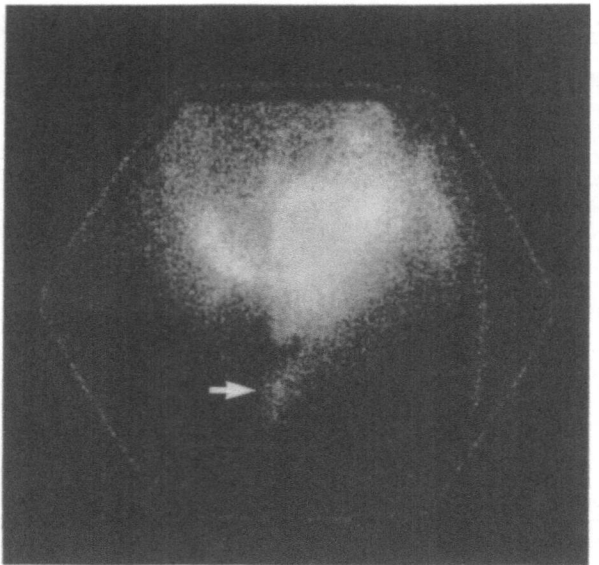

This patient on CAPD had labial edema but no structural defect was present on physical exam. The peritoneoscintigraphy performed 5 hr after the administration ^{99m}Tc SC in the dialysate demonstrated radioactivity in the left labia (*arrow*), which was compatible with a left patent processus vaginalis.

Comment: This case demonstrates an additional site of edema with subsequent identification of the structural defect, which was not noted clinically. Because of increased sustained pressure, CAPD may aggravate a preexisting or create a new structural defect, which may then leak dialysate resulting in edema (e.g., labial, scrotal, penile, buttocks, and thighs). Peritoneoscintigraphy appears to be most useful in the evaluation of edema when no structural defect is present clinically.

Radioactivity at the site of leakage may be better or only identified on delayed images.

(Reproduced with permission from Kopecky RT, Frymoyer PA, Witanowski LS, et al. Complications of continuous ambulatory peritoneal dialysis: diagnostic value of peritoneal scintigraphy. *Am J Kidney Dis.* 1987;10:123.)[4]

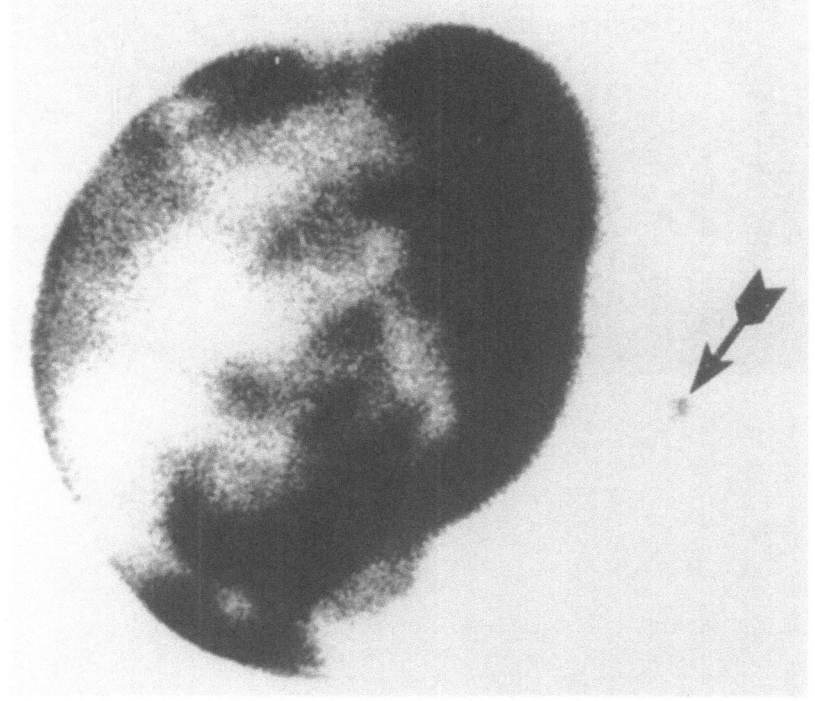

A

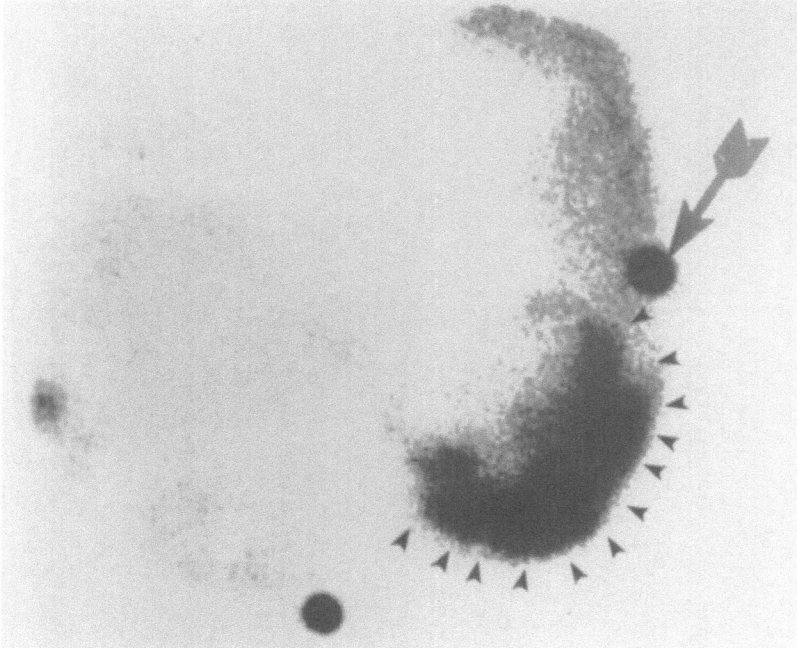

B

Figure 5.20. CAPD: value of post drainage images.

The anterior image (**A**), which was performed after the administration of 5 mCi of ^{99m}Tc SC in dialysate fluid, demonstrates apparently normal distribution. However, a repeat anterior image (**B**) was performed after drainage of the dialysate and demonstrates an abnormal persistence (*arrowheads*) of radioactivity corresponding to swelling and the patient's complaints in the left lower quadrant of the abdomen. This confirmed the presences of leakage into the abdominal wall, which was successfully repaired. A marker (*arrow*) denotes the catheter exit site; a second marker demonstrates the symphysis pubis.

Comment: This case emphasizes the importance of postdrainage images.

(Reproduced from Berman C, Velchik MG, Shusterman N, et. al. The clinical utility of the Tc-99m SC intraperitoneal scan in CAPD patients. *Clin Nuc Med*. 1989;14: 405.)[5]

Figure 5.21. CAPD: retained lesser perito-neal sac dialysate.

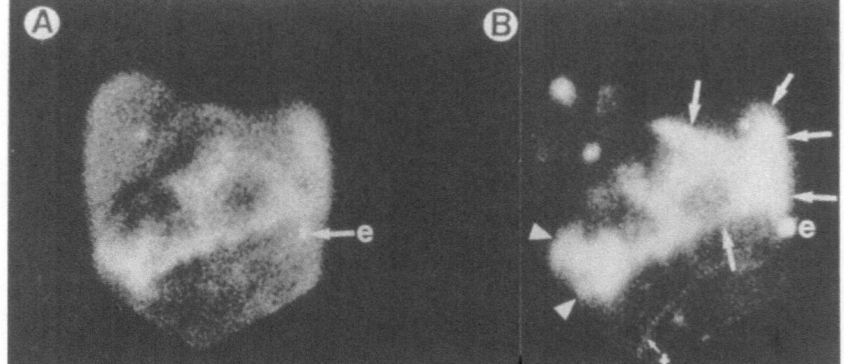

This patient observed persistent fullness and intermittent poor catheter drainage of the dialysate fluid after CAPD. Anterior ^{99m}Tc SC peritoneoscintigraphy at 2 hr postinjection (**A**) demonstrated apparently good distribution of radioactivity in the peritoneal cavity with a suggestion of in-creased radioactivity in the midabdomen. The delayed anterior image (**B**) performed after drainage of the dialysate fluid demon-strated retained radioactivity in Morrison's pouch (*arrowheads*) and the lesser perito-neal sac (*arrows*). As confirmed at initial dialysis catheter insertion, the patient had multiple postoperative adhesions within the great peritoneal sac. These may have re-sulted in significant obliteration of the greater peritoneal sac with subsequent visu-alization of the lesser sac. The catheter exit site is marked as "e," and the external dialy-sate tubing is marked as "t."

Comment: This case demonstrates the pat-tern of visualization of the lesser sac and the value of peritoneoscintigraphy in diag-nosing the etiology of poor dialysate drain-age. This case again emphasizes the value of delayed post–dialysate-drainage images. (Reproduced with permission from Ko-pecky RT, Frymoyer PA, Witanowski LS, et. al. Prospective peritoneal scintigraphy in patients beginning continuous ambula-tory peritoneal dialysis. *Am J Kidney Dis.* 1990;15:228–236.)[11]

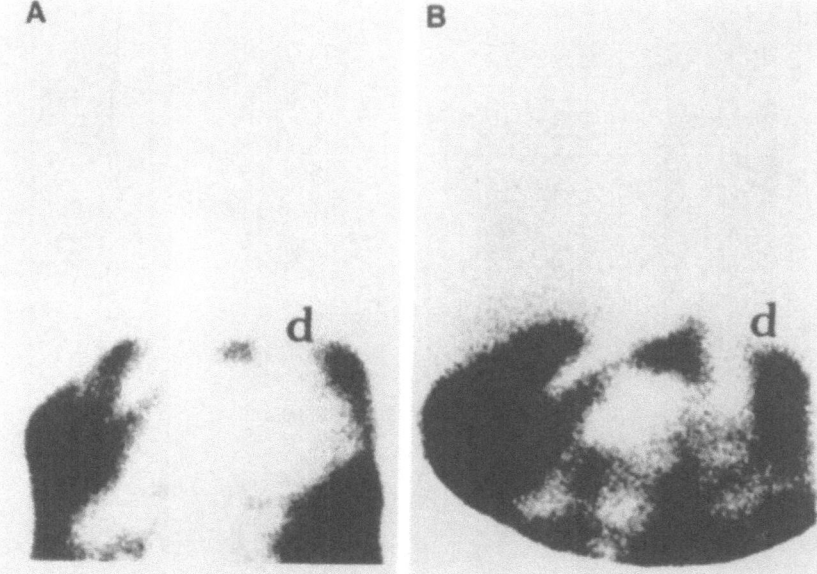

Figure 5.22. CAPD: transdiaphragmatic leakage.

This patient on CAPD developed exertional shortness of breath and right pleuritic chest pain. Chest radiograph revealed right pleural effusion. However, evaluation for etiology including ventilation/perfusion scan, cytologic examination of fluid, and methylene blue instillation in the dialysate with subsequent reaspiration of the pleural effusion was unremarkable. Peritoneoscintigraphy with additional images of the thorax was performed with 5 mCi of ^{99m}Tc albumin colloid. The anterior image of the upper abdomen and thorax (A) demonstrated very faint radioactivity over the right hemithorax, which did not photograph well; the anterior image (B), which was performed at 2 hr, demonstrated definite radioactivity in the left hemithorax confirming a transdiaphragmatic leak. (The level of the diaphragm is represented by "d.")

Comment: Although transdiaphragmatic leak is an unusual etiology of pleural effusion, in those patients with a pleural effusion of undetermined etiology and a predisposition for transdiaphragmatic leak (e.g., CAPD, ascites secondary to cirrhosis, etc.), peritoneoscintigraphy may be of utility in noninvasively confirming the etiology. Delayed imaging even on the following day may be essential to make the diagnosis.

(Reproduced with permission from Walker JV, Fish MB. Scintigraphic detection of abdominal wall and diaphragmatic peritoneal dialysis in patients on continuous ambulatory peritoneal dialysis. *J Nucl Med*. 1988; 29:1596.)[12]

Figure 5.23. Monoclonal antibody imaging (normal).

The ^{131}I labeled B72.3 monoclonal antibody (^{131}I B72.3) has been used in the evaluation of peritoneal metastasis from colorectal and ovarian carcinoma. In this case, the ^{131}I B72.3 has been injected intraperitoneally with normal distribution noted on the immediate image. Over 7 days, decreasing but normal activity is noted in the abdomen with no abnormal sites of ^{131}I B72.3 noted to suggest peritoneal metastasis. The radioactivity lateral and outside of the left lower abdomen on the 3-day image represents contamination in the gauze covering the Tenckhoff catheter exit site. The radioactivity in the lower pelvis on the 1-, 3-, and 7-day images represents normal bladder activity. The radioactivity in the neck region on the 7-day image represents thyroid. A marker is present in the right upper corner of the images, which becomes more prominent as longer imaging time and intensity are used.

Comment: The normal distribution of an intraperitoneal injected radiopharmaceutical has been described in the initial cases of this atlas. This case demonstrates the normal clearance and thus the normal distribution over time of the intraperitoneal injected monoclonal antibody ^{131}I B72.3. This case also demonstrates the normal organs or artifacts which may be visualized at different times with this radiolabeled monoclonal antibody. The various organs visualized may vary with the pharmacokinetics of the absorbed radiopharmaceutical, its breakdown products, and/or metabolites. (Reproduced with permission from Carrasquillo JA, Sugarbaker P, Colcher D, et. al. Peritoneal carcinomatosis: imaging with intraperitoneal injection of ^{131}I-labeled B72.3 monoclonal antibody. *Radiology.* 1988;167:35–40.)[13]

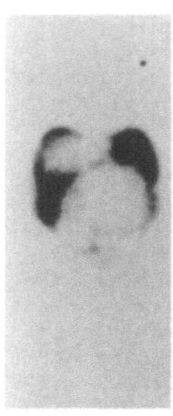 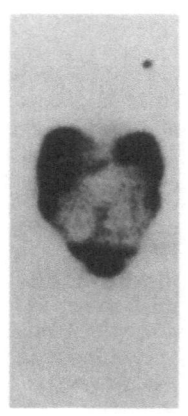 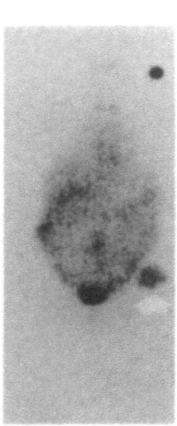 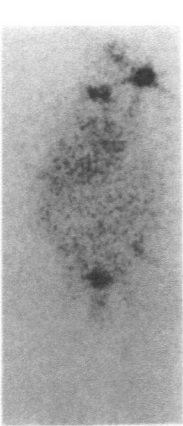

Immediate   1 day   3 days   7 days

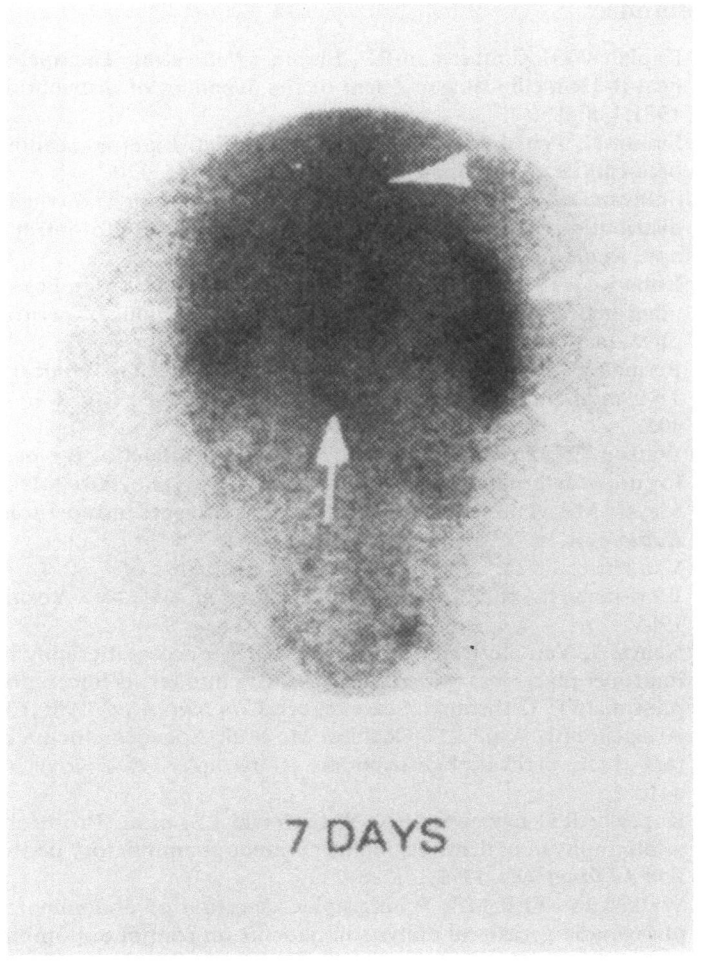

7 DAYS

Figure 5.24. Monoclonal antibody imaging (abnormal).

In a 33-year-old woman with a ruptured appendiceal adenocarcinoma, [131]I B72.3 was injected intraperitoneally. The immediate image demonstrated inhomogeneous peritoneal radioactivity, which was within the limits of normal. However, on the 7-day image, focal areas of radioactivity are present and represent residual tumor overlying and adjacent to the liver and residual tumor encasing the spleen (*arrows*). The CT scan showed no evidence of residual tumor.

Comment: This case demonstrates the outstanding potential of monoclonal antibody imaging and the complementary role to anatomical imaging (e.g., CT).

(Reproduced with permission from Carrasquillo JA, Sugarbaker P, Colcher D, et. al. Peritoneal carcinomatosis: imaging with intraperitoneal injection of I-131-labeled B72.3 monoclonal antibody. *Radiology*. 1988;167:35–40.)[13]

References

1. Kaplan WD, Zimmerman RE, Bloomer WD, et al. Therapeutic intraperitoneal P-32: a clinical assessment of the dynamics of distribution. *Radiology.* 1981;138:683.

2. Simon N, Feitelbergh S, Warner RRP, et al. External scanning of internal beta-emitters. *J Mt Sinai Hosp NY.* 1966;33:365–370.

3. Sullivan DC, Harris CC, Currie JL, et al. Observations on the intraperitoneal distribution of chromic phosphate (P-32) suspension for intraperitoneal therapy. *Radiology.* 1983;146:539.

4. Kopecky RT, Frymoyer PA, Witanowski LS, et al. Complications of continuous ambulatory peritoneal dialysis: diagnostic value of peritoneal scintigraphy. *Am J Kidney Dis.* 1987;10:123.

5. Berman C, Velchik MG, Shusterman N, et al. The clinical utility of the Tc-99m SC intraperitoneal scan in CAPD patients. *Clin Nuc Med.* 1989;14:405.

6. Watson EE, Stabin MG, Davis JL, et al. A model of the peritoneal cavity for use of internal dosimetry. *J Nucl Med.* 1989;30:2002–2011.

7. Meyers MA. The spread and localizations of acute intraperitoneal effusions. *Radiology.* 1970;95:547–554.

8. Van Nostrand D, Silberstein EB. Therapeutic use of P-32. In: Freeman LM, Weissman HS, eds. *Nuclear Medicine Annual 1985.* New York: Raven Press; 1985.

9. Neutze J, Van Nostrand D, Major W. Peritoneoscintigraphy in detection of improper placement of peritoneal catheter into bowel lumen prior to chromic phosphate P-32 therapy: a case report. *Clin Nucl Med.* 1985;10:777–779.

10. Arnstein NB, Wahl RL, Cochran M, et al. Adenocarcinoma of the alimentary tract: peritoneal distribution scintigraphy. *Radiology.* 1987;162:439–441.

11. Kopecky RT, Frymoyer PA, Witanowski LS, et al. Prospective peritoneal scintigraphy in patients beginning continuous ambulatory peritoneal dialysis. *Am J Kidney Dis.* 1990;15:228–236.

12. Walker JV, Fish MB. Scintigraphic detection of abdominal wall and diaphragmatic peritoneal dialysis in patients on continuous ambulatory peritoneal dialysis. *J Nucl Med.* 1988;29:1596.

13. Carrasquillo JA, Sugarbaker P, Colcher D, et al. Peritoneal carcinomatosis: imaging with intraperitoneal injection of I-131-labeled B72.3 monoclonal antibody. *Radiology.* 1988;167:35–40.

Index